THE BIG NECESSITY

ROSE GEORGE has written for the *Guardian*, the *Independent*, the *London Review of Books*, *Jane* and many other publications. She graduated with first class honours from the University of Oxford, and also holds a master's degree in international relations from the University of Pennsylvania. Her writing assignments have included two trips to Saddam Hussein's birthday party in Tikrit, and one to Bhutan to watch the Alternative World Cup soccer final between Bhutan and Montserrat. Her first book, *A Life Removed: Hunting for Refuge in the Modern World*, about Liberian refugees, was longlisted for the Lettre Ulysses Award for reportage. She lives in London.

The Big Necessity

Adventures in the World of Human Waste

Rose George

Published by Portobello Books Ltd 2008

Portobello Books Ltd
Twelve Addison Avenue
Holland Park
London
W11 4QR

A CIP catalogue record is available from the British Library

9 8 7 6 5 4 3 2 1

ISBN 978 1 84627 069 7

www.portobellobooks.com

Text designed and typeset in Berkeley Book by
Avon DataSet Ltd, Bidford on Avon, Warwickshire

Printed in the UK by CPI William Clowes Beccles NR34 7TL

For T.D.W. George, for the introduction.

Contents

Like an apartment where there's kitchen, bedroom and bathroom. People see that and they want the same for themselves, a bigger house with different rooms for everything. They can't have all that so they get the big necessity, a toilet.

Sheikh Razak, slum toilet builder, Mumbai

Chengdu, China – *Author*

INTRODUCTION

Examining the Unmentionables

I need the toilet. I assume there is one, though I'm at a spartan restaurant in the Ivory Coast, in a small town filled with refugees from next-door Liberia, where water comes in buckets and you can buy towels second-hand. The waiter, a young Liberian man, only nods when I ask. He takes me off into the darkness to a one-room building, switches on the light and leaves. There's a white tiled floor, white tiled walls and that's it. No toilet, no hole, no clue. I go outside to find him again and ask if he's sent me to the right place. He smiles with sarcasm. Refugees don't have much fun but he's having some now. 'Do it on the floor. What do you expect? This isn't America!' I feel foolish. I say I'm happy to use the bushes, it's not that I'm fussy. But he's already gone, laughing into the darkness.

I need the toilet. I leave the reading room of the British Library in central London and find a 'ladies' a few yards away. If I prefer, there's another one on the far side of the same floor, and more on the other five floors. By 6 p.m., after thousands of people have entered and exited the library and the toilets, the stalls are still clean. The doors still lock. There is warm water in the clean sinks. I do what I have to do, then flush the toilet and forget it, immediately, because I can, and because all my life I have done no differently.

This is why the Liberian waiter laughed at me. He thought that I thought a toilet was my right, when he knew it was a privilege.

It must be, when 2.6 billion people don't have sanitation. I don't mean that they have no toilet in their house and must use a public one with queues and fees. Or that they have an outhouse, or a rickety shack that empties into a filthy drain or pigsty. All that counts as sanitation, though not a safe variety. The people who have those are the fortunate ones. Four in ten people have no access to any latrine, toilet, bucket or box. Nothing. Instead, they defecate by train tracks and in forests. They do it in plastic bags and fling them through the air in narrow slum alleyways. If they are women, they get up at 4 a.m. to be able to do their business under cover of darkness for reasons of modesty, risking rape and snakebites. Four in ten people live in situations where they are surrounded by human excrement, because it is in the bushes outside the village, or in their city yards, left by children outside the back door. It is tramped back in on their feet, carried on fingers onto clothes, food and drinking water.

The disease toll of this is stunning. A gram of faeces can contain 10 million viruses, 1 million bacteria, 1000 parasite cysts, and 100 worm eggs. Bacteria can be beneficial: the human body needs bacteria to function, and only ten per cent of cells in our body are actually human. But plenty are malign. Small faecal particles can contaminate water, food, cutlery and shoes, and be ingested, drunk or unwittingly eaten. One sanitation specialist has estimated that people who live in areas with inadequate sanitation ingest ten grams of faecal matter every day. Poor sanitation, bad hygiene and unsafe water – usually unsafe because it has faecal particles in it – cause one in ten of the world's illnesses. Children suffer most. Diarrhoea – nearly ninety per cent of which is caused by faecally contaminated food or water – kills a child every fifteen seconds. The number of children dead from diarrhoea over the last decade exceeds all people killed by armed conflict since the Second World War. Diarrhoea, says the UN children's agency UNICEF, is the largest hurdle a small child in a developing country must overcome. Larger than AIDS, or TB, or malaria. It is estimated that 2.2 million people – mostly children – die from an affliction that to most Westerners is the result of a bad take-away. Public health professionals talk about water-related diseases, but that is a euphemism for the truth. These are shit-related diseases.

In 2007, readers of the *British Medical Journal* were asked to vote for the biggest medical milestone of the last 200 years. Their choice was wide: antibiotics, penicillin, anaesthesia, the Pill. They chose sanitation. In poorly sewered nineteenth-century London, one child out of two died early. After toilets, sewers and hand-washing with soap became normal, child mortality dropped by a fifth. It was the largest reduction in child mortality in British history. In the poor world, proper disposal of human excreta – the process which is given the modern euphemism of 'sanitation' – can reduce diarrhoea by nearly forty per cent. (Though ninety per cent of most sanitation-related budgets go on water supply, providing more or cleaner water only reduces diarrhoea by sixteen to twenty per cent.) Harvard University geneticist Gary Ruvkun believes the toilet is the single biggest variable in increasing human life-span. Modern sanitation has added twenty years to the average human life. Good sanitation is also economically sensible. A government that provides adequate sanitation saves money on hospital visits avoided, does not lose labour days to dysentery or workers to cholera. Where good sanitation exists, people are wealthier, healthier and cleaner.

When sanitarians talk about history, their timeline usually begins on a Friday morning in 1854, when Dr John Snow, a doctor in Soho, removed the handle from a water pump in Broad Street, because he was the first to understand that cholera was travelling in excrement that got into the water supply, and the awful consequences of this fact (in 1849, cholera killed over 50,000 people nationwide). Sewers followed; flush toilets flourished. By now, modern living provides nearly everyone with one or several magic disposal units that make excrement disappear and that act as a barrier between humans and their potentially toxic waste. Every city has sewers that take it away to who knows where, where a bigger disposal unit does with it whatever it does, out of sight and hopefully without smell. Sanitation is what modern cities are built on and how they can function with so many people living so closely together, without the consequences that the plastic-bag defecators know too well, because it cripples their guts and kills their children.

Rich toileted people; poor toiletless masses. Life, luxury and health for the privileged. Disease and death and business as usual for the poor. This is the assumption the Liberian waiter relied on to make me feel embarrassed. He was entitled to it, because he was a refugee, and diarrhoea probably kills more refugees – in camps, on the run – than soldiers or guerrillas. But he was mistaken.

In the spring of 2007, the city of Galway, on the west coast of Ireland, held its annual arts parade. Galway has a reputation as a cultural centre. It has a good university. It has nice benches in pleasant parks, including one on which I once sat and watched, dumbly, as a hand snaked over and stole my backpack, then listened as a crowd of shouting men immediately spilled out of a nearby pub and set off in pursuit of the thief out of the goodness of their Guinness-filled hearts.

I have good memories of Galway, but I'm glad I wasn't there that year, because the newest addition to the Galway Arts Parade was a man in a green fuzzy costume with many arms and one eye. He had been given the name Crypto, and anyone who had been in Galway for the previous five months would have needed no further introduction, because Crypto was the reason that a world-class cultural city was living with the conditions familiar to any inhabitant of the world's worst slums. Crypto was a big cuddly version of a parasite called cryptosporidium, a disease-causing protozoa – a single-celled, amoeba-like organism – that can travel in faeces. For five months and counting, Crypto and his billions of cousins had been the reason that a rich and developed city in a rich and developed country had no drinking water. A cultural centre of Europe, in a land wealthy enough to be nicknamed the Celtic Tiger, was forced to issue boil-water notices more familiar in places of poverty and dust where children die young.

It had begun in early March, with reports of persistent stomach aches and diarrhoea. There were hospitalizations, of the vulnerable (the old, the young, the immuno-suppressed) and there was bewilderment as to the cause. Something had polluted the drinking-water supply of Lough Corrib. First the cows were blamed. They must have been defecating nearby. Then it was the farmers; run-off from their pesticide-treated fields could have polluted the water. Then someone began to suspect sewage. Initial tests found that most infections were caused by cryptosporidium hominis, which passes from human to human. An investigative programme on Ireland's national radio station found that levels of cryptosporidium in effluent discharged into the lough from Oughterard sewage works were 600 times levels permitted in neighbouring Northern Ireland.

Two facts about Galway's cryptosporidium crisis held my attention. First, that the scandal didn't reverberate beyond Ireland's borders, though an advanced society that had supposedly known how to dispose of its sewage for nearly 150 years was suddenly unable to provide its citizens with water uncontaminated by faeces. Second, that it was not unexpected. After the outbreak was publicized, the people of Ennis, the seat of County Clare, did some crypto one-upmanship. You may have had no water for five months, they said, but we haven't had any for two years (and they won't have any until 2009, when a water treatment plant should be completed). A fifth of Ireland's towns are at high risk of crytosporidium infection, according to the national environmental protection authority. Nearly half the country treats its sewage only to primary levels, which involves nothing more taxing than screening out the lumps and discharging the rest. And Ireland is not the only rich country with an infrastructure more suited to a poor one.

Milan, Italy's cultural capital, has a world-class opera house, La Scala, and is an international fashion capital, but until shamefully recently, it couldn't manage to do anything with its sewage but discharge it raw and dangerous into the suffering river Lambro. The city finally built its first treatment plant three years ago, possibly spurred by a threat from the European Union of being fined $15 million a day for contravening a waste disposal directive. This is ironic, considering that Brussels, the wealthy and powerful city that serves as the EU's administrative seat, only began to build a treatment plant for its own sewage in 2003. Before that, it sent the waste of all those diplomats, bureaucrats and clever, competent people into a river, and those clever, competent people didn't question it. In the United States, cryptosporidium in Milwaukee's drinking water made 400,000 people sick and killed more than 100. It was the biggest water-contamination disease outbreak in US history, and it happened in 1993, over a century after the city fathers of America installed pipes to bring clean water to their citizens, and sewers and treatment plants to take the foul water away. But to where? Milwaukee discharges treated sewage effluent – treated to remove some things, but not pharmaceuticals or all pathogens – into Lake Michigan, which also supplies its drinking water. Sometimes it discharges raw sewage too. Since 1994, 935 million gallons of 'full-strength, untreated sewage' have been poured into the lake's waters. This is not illegal. In fact, it's what the system is designed to do if too much storm-water overloads storage capacity at treatment works.

Ninety per cent of the world's sewage ends up untreated in oceans, rivers and lakes, and a fair share comes from the sanitary cities supplied with sewers and treatment plants. Sanitation in the Western world is built with pipes but on presumption. Despite the technology, the engineers and the ingenuity of modern sanitary systems, despite the shine of progress and flush toilets, even the richest, best-equipped humans still don't know what to do with sewage except move it somewhere else and hope no-one notices when it is poured untreated into drinking-water sources. And they don't.

In 2006, I wrote a series on sewage for the online magazine *Slate*. Later, I found a comment that a reader left on the discussion page. 'Someone at *Slate* a scat freak? Is this all some giant experiment to see if we have no sense of class or dignity?' It made me smile but it wasn't surprising. I've spent many months now answering the question of why I'm writing a book like this. The interrogation happens so often, my responses have become routine.

First I establish that I am no scatologist, fetishist or coprophagist. I don't much like toilet humour (and by now I've heard a lot of it). I don't think 2.6 billion people without a toilet is funny. Then, I tailor my answers and language to the social situation – still managing to spoil many lunches – and by explaining the obvious. Everyone does it. It's as natural as breathing. The average human being spends three years of life going to the toilet, though the average human being with no physical toilet to go to probably does his or her best to spend less. It is a human behaviour as revealing as any other about human nature, but only if it can be released from the social straitjacket of denial. Rules governing defecation, hygiene and pollution exist in every culture at every period in history.

It may in fact be the foundation of civilization: what is toilet training if not the first attempt to turn a child into an acceptable member of society? Appropriateness and propriety begin with a potty. How a society disposes of its human excrement is an indication of how it treats its humans too. Unlike other body-related functions like dance, drama and songs, wrote the Indian sanitarian Dr Bindeshwar Pathak, 'defecation is very lowly'. Yet when discussing it, he continued, 'one ends up discussing the whole spectrum of human behaviour, national economy, politics, role of media, cultural preference and so forth'. And that's a partial list. It is missing biology, psychology, chemistry, language. It is missing everything that touches upon understanding what the American development academic William Cummings called 'the lonely bewilderment of bodily functions'.

If my questioner is religious, I say that all the world's great faiths instruct their followers how best to manage their excrement, because hygiene is holy. I explain that taking an interest in the culture of sanitation puts them in good company. Mohandas K. Gandhi, though he spent his life trying to rid India of its colonial rulers, nonetheless declared that sanitation was more important than independence. The great architect Le Corbusier considered the toilet to be 'one of the most beautiful objects industry has ever invented' and Rudyard Kipling found sewers more compelling than literature. Drains are 'a great and glorious thing,' he wrote in 1886, 'and I study 'em and write about 'em when I can.' A decent primer on sanitary engineering, he wrote, 'is worth more than all the tomes of sacred smut ever produced'. Anton Chekhov was moved to write about the dreadful sanitation in the Far-Eastern Russian isle of Sakhalin. And Sigmund Freud thought the study of excretion essential and its neglect a stupidity. In the foreword to *Scatologic Rites of All Nations*, an impressive ethnography of excrement by the amateur anthropologist and US army captain John Bourke, Freud wrote that 'to make [the role of excretions in human life] more accessible... is not only a courageous but also a meritorious undertaking'.

Solving sanitation is also a noble pursuit, if the amount of royals who are interested in it is an indication: Prince Charles of the House of Windsor cleans his wastewater naturally by sending it slowly through a pond filled with reeds. King Bhumibol of Thailand holds a patent for a wastewater aerator, making him the only patent-holding monarch in the world. Prince Willem-Alexander of Orange, heir to the Dutch throne, leads the UN's sanitation advisory body. It takes a brave academic to address it, but the ones who do rise to the occasion, producing papers like *My baby doesn't smell as bad as yours: The plasticity of disgust*, by the psychologists Trevor Case, Betty Repacholi and Richard Stevenson; or *The Scatological Rites of Burglars* by Albert B. Friedman, a noted professor of medieval literature, who must have been tickled to learn that the housebreaker's habit of leaving a foul deposit is probably an ancient custom, and was alluded to in seventeenth-century German literature.

If the cultural standing of excrement doesn't convince them, I say that the material itself is as rich as oil and probably more useful. It contains nitrogen and phosphates which can make plants grow and also suck the life from water, because its nutrients absorb available oxygen. It can be both food and poison. It can contaminate and cultivate. Millions of people cook with gas made by fermenting it. I tell them I don't like to call it 'waste', when it can be turned into bricks, when it can make roads or jewellery, and when in a dried powdered form known as *poudrette*, it was sniffed like snuff by the grandest ladies of the eighteenth-century French court. Medical men of not too long ago thought stool examination a vital diagnostic tool (London's Wellcome Library holds a 150-year-old engraving of a doctor examining a bedpan and a sarcastic maid asking him if he'd like a fork). They were also fond of prescribing it: excrement could be eaten, drunk or liberally applied to the skin. Martin Luther was convinced: he reportedly ate a spoonful of his own excrement daily, and wrote that he couldn't understand the generosity of a God who freely gave such important and useful remedies.

This may seem like quackery, except that the faecal transplant is becoming an increasingly common procedure in modern medicine, used to treat severe bacterial infections such as clostridium difficile, known by tabloids as a 'superbug' because of its resistance to many antibiotic remedies. For the worst-suffering cases, doctors can now prescribe an enema – mixed with milk or saline solution – of a close relative's disease-free faeces, whose bacterial fauna somehow defeat the superbug with dramatic effect. (Ninety per cent of patients given faecal transfusions recover.) An eighty-three-year-old Scottish granny called Ethel McEwen, freshly cured by a dose of her daughter's faeces, said it wasn't much different from a blood or kidney transplant, and anyway, 'it's not like they put it on a plate and have you eat it. You don't ever see or smell a thing.'

My sales technique nearly always worked. One evening over beer, an Indian novelist asked with seemingly bored politeness what I was working on, then talked for an hour of New Zealand 'long-drops' (deep pit latrines) and whether it is acceptable to answer the phone while on the toilet, a modern question of etiquette that defeats me. My neighbour's elderly mother reminisced about the outdoor privy she had as a child, and about the man who called to collect the urine, which he then sold to tanners, and she sounded like she misses both. Pub conversations regularly took a toilet turn: one regular greeted me one day by saying that he only ever urinated sitting down. An expression of relief crossed his face, before he turned back to his pint.

To research *The Bathroom*, an exhaustive exploration of human toilet habits, the architect Alexander Kira surveyed 1000 Californians. In an article he wrote for *Time* headlined 'Examining the Unmentionables', Kira said, 'Once people got talking about bathrooms they couldn't stop.'

The toilet is a physical barrier that takes care of the physical dangers of excrement. Language takes care of the social ones. In *The Civilizing Process*, the anthropologist Norbert Elias charts the progression of human defecation from a public, unremarkable activity – it was considered an honour to attend monarchs seated on their commodes – to a private, shameful one, done behind closed doors and, except in China, never in company.

Newspapers are fond of anointing last taboos, but in modern, civilized times, the defecatory practice of humans is undeniably a candidate. Sex can be talked about, probably because it usually requires company. Death has once again become conversational, enough to be given starring roles in smart, prime-time TV dramas. Yet defecation remains closed behind the words, all chosen for their clean association, that we now use to keep the most animal aspect of our bodies in the backyards of our discourse where modernity has decided it belongs. Water closet. Bathroom. Restroom. Lavatory. Sometimes, we add more barriers by borrowing from other people's languages. The English took the French *toilette* (a cloth), and used it first to describe a cover for a dressing table, then a dressing room, then the articles used in the dressing room, and finally, but only in the nineteenth century, a place where washing and dressing was done, and then neither washing nor dressing. (They also borrowed *Gardez l'eau*, commonly shouted out before throwing the contents of chamber pots into the streets, and turned it into 'loo'.) The French, in return, began by calling their places of defecation 'English places' (*lieux à l'anglaise*), and then took the English acronym WC instead. The Japanese have dozens of native words for a place of defecation, but prefer the Japanese-English *toiretto*. You have to go back to the Middle Ages to find places of defecation given more accurate and poetic names: many a monk used a 'necessary house'. Henry VIII installed a House of Easement at his Hampton Court Palace. The easiest modern short-hand for the disposal of human excreta – sanitation – is a euphemism for defecation which is a euphemism for excretion which is a euphemism for shitting. This is why the young boy hero of Dr Seuss' *It's Grinch Night* can ask for permission 'to go to the euphemism'. This is why the only safe place for modern humans to talk about defecation is in the unthreatening embrace of humour, and why the ordinary, basic activity of excretion has been invested with an emotional power that has turned a natural function into one of our strongest taboo words.

Our disgust with shit seems deep and sure, as potent as the swear words which get their power from it. There are good biological reasons for this. Faeces are unpleasant. Outside the sexual fetish world of coprophagy, no-one wants to smell, feel or touch them (including me). But the power of our taboo words is modern. Church words used to hurt much more. The diminished power of 'damn' explains why the climax of *Gone with the Wind* is always a bit of a puzzle. When church influence weakened, the products of the body – which Puritan influence has successfully turned into a foul, shameful thing – stepped in instead to give us our worst words. There must be something wrong with it, after all, when all we do is get rid of it as fast as possible.

Meanwhile, a plentiful supply of euphemisms can serve as linguistic stand-ins. The cognitive scientist Steven Pinker lists a dozen categories of euphemism, including taboo ('shit'), medical ('stool', 'bowel movement') and formal ('faeces', 'excrement', 'excreta', 'defecation', 'ordure'). The category that is missing is 'conversational'. There is no neutral word for what humans produce at least once a day, usually unfailingly. There is no defecatory equivalent of the inoffensive, neutral 'sex'.

I wish that 'shit' didn't shock. It is a word with noble roots, coming from a family of words that also contains the Greek *skihzein*, the Latin *scindere* or the Old English *scitan*, all meaning, sooner or later, to divide or separate. (Science is the art of distinguishing things by knowledge.) I use it, sometimes, because of a frustration with all those euphemisms. Faeces is the Latin word for dregs and only took on its modern meaning in the seventeenth century. Any proponent of ecological sanitation – the re-use, via composting or some other means, of human excrement – will object to a potentially powerful and inexhaustible fertilizer being thought 'the most worthless parts'. They also object to 'waste', because it derives from the Latin for 'uncultivated', and because it shouldn't be wasted.

But mostly I use the word because throughout my travels the people who deal with things best are the ones who are not afraid of it. In the words of Umesh Panday, a Nepali sanitation activist, 'just as HIV/AIDS cannot be discussed without talking frankly about sex, so the problem of sanitation cannot be discussed without talking frankly about shit'.

One evening in Bangkok, I attend a party. It's the end of a long day, and the Toilet Party is supposed to calm spirits and foster connections, because it is being held as part of the World Toilet Organization's 2006 Expo. The Thais treat the conference with respect, perhaps because of their beloved king's interest in wastewater aerators: a couple of hundred Thai delegates have been summoned from various government departments, from all over the country.

On the stage, traditional Thai dancers, with tapering fingers and extreme beauty, manage to glide with a serenity undiminished by the toilets that serve as a backdrop. As the entertainment proceeds, the attendees mingle over buffet food. A Japanese man who speaks no English, but who proffers a business card declaring himself a 'household paper historian', tries to converse with a world authority on public toilets, who speaks no Japanese. The man in charge of Bangkok's sewage disposal has an earnest discussion with a Sri Lankan who has spent two years building low-cost latrines for tsunami victims. A man with a moustache introduces himself as a TV star in Malaysia. He's a TV cop. 'Actually, the Malaysian equivalent of Starsky, as in Hutch.' Here, over canapés, is everything that intrigues me about this hidden human activity. Dedicated people, derided but determined, toasting their unheralded efforts to solve the world's biggest unsolved public health crisis, because who else, outside this world, will do it for them?

By the time of the party, I have had a professional curiosity in human excreta disposal for several months. I've noticed something strange happening. I read a piece about the Austrian director Michael Haneke in the *New York Times* and he suddenly says, out of nowhere, 'We have a saying in Austria: we are already up to our necks in sewage. Let's not make waves.' I turn on the TV with no programme in mind and find a documentary about W.H. Auden, and the first talking head is saying how Auden's guests were only allowed one sheet of toilet paper because any more was wasteful (I liked Auden already; I like him more now I know this). On a train travelling through France one day, I become immersed in a book for hours. The first time I look up and out of the window, I see a sewage treatment plant in the middle of a green field. Psychologists call this a perception filter. Once you notice something, you notice it everywhere. Our most basic bodily function, and how we choose to deal with it, leaves its signs everywhere entwined with everything, as intricately intimate with human life as sewers are with the city. Under our feet, at the edge of sight, but there.

Once I start noticing, I can't stop. And once I start meeting the people who work in this world – who flush its sewers and build its pit latrines; who invent and engineer around our essential essence, in silence and disregard – I don't want to. I'd rather follow Sigmund Freud, who wrote that humanity's 'wiser course would undoubtedly have been to admit [shit's] existence and dignify it as much as nature will allow'. So here goes.

East 21st Street, between Broadway and Park Avenue – *New York City Department of Environmental Protection*

1. IN THE SEWERS

The Art of Drainage

Beside a manhole in an East London street, a man called Happy hands over the things that will protect me in the hours to come. White paper overalls, with hood. Crotch-high waders with tungsten-studded soles that will grip but won't spark. A hard hat with a miner's light. Heavy rubber gloves, oversized. A 'turtle' – a curved metal box containing an emergency breathing apparatus – to strap around my waist, along with a back-up battery. Finally, a safety harness that Happy helps me buckle with delicacy, as it loops through my legs near my groin. It's tight but comfortable, and has the side benefit for male wearers of making all men seem rather well-endowed. The harness will be the only means of dragging me out from the sewer I am about to descend into, where the hazards include bacteria and viruses such as hepatitis A and B and C, rabies and typhoid, and leptospirosis ('sewer workers' disease') that can be caught from rat urine, and in its severe form causes vomiting, jaundice and death.

There are also the gases. Methane, obviously. Hydrogen sulphide, known as sewer gas, which forms when organic matter decomposes in sewage, smells like rotten eggs and kills by asphyxiation. And whatever fumes arise from whichever effluents London's commercial businesses have chosen to pour down their drains and toilets today, with proper warning or not. The greatest danger is the flow, which can be increased suddenly and rapidly with rainfall, so a stream becomes a torrent, and one that can contain anything that has been put down it that day, from two-by-fours to pieces of 4x4s. Sewer workers have always died on the job, and they still die, no matter how advanced the infrastructure. In March 2006, Minnesota workers Joe Harlow and Dave Yasis drowned in the St Paul sewer system when a rainstorm came on suddenly.

Water can be dangerous in other ways: sugar manufacturers, for example, send into the sewer the boiling water that they clean their vats with. Underground, it turns into steam and can react badly with other gases in the system. Sewers that are known to be especially hazardous are ranked C-class, and cannot be entered without special permits. Though the men accompanying me have worked in the sewers for decades, they cannot know every inch of a vast network nor what is likely to be discharged into it. Some sewers haven't been visited in fifteen years. It's best to be prepared. And indemnified: a paper-suited man thrusts a form at me, as I struggle with my crotch-high waders (items of clothing that would make members of the online Yahoo sewer-boots fetish group – which does exist – speechless with one emotion or another). He says, 'Sign this,' and gives me no time to read it. 'Don't worry,' he says, with no smile. 'It just means if you collapse, I get all your money.' This is flusher humour. It helps in a hard job, and there will be more of it.

Half a dozen men stand around the manhole. They match well enough the London journalist Henry Mayhew's description of their predecessors in 1851: 'Well-conducted men generally, and for the most part, fine stalwart good-looking specimens of the English labourer', though the size of their paunches shows that they've moved on from the traditional sewerman's tipple of rum to beer. They all seem to have very white teeth.

My escorts include one consultant, one senior engineer and several wastewater operatives. Their names are Dave and Keith and Rob and Happy, but in the language of those who work in the city's sewers, they're all flushers. The name is no longer used officially, because it describes the job in times past when men waded into the silt of a sewer and dislodged blockages with brooms and rakes, and opened inlets to flush river water into the tunnels to nudge the flow down into the Thames. They're wastewater operatives now, but they do what the flushers did: they keep the flow flowing.

Their equipment is better than the heavy blue overcoats and wick lamps that flushers used a century ago, but the men are fewer. If you look at the sewer systems of great cities, you'll start to think there's something wrong with the maths. New York's 6000 miles of sewers are served by 300 flushers. Paris has 1500 miles and 284 *égoutiers*. The mightiest network of any metropolitan city is London's. It is so mighty, no-one knows how big it is. Thames Water, the private water utility that serves London, has 37,000 miles of sewers in its whole catchment, but that extends 80 miles from Central London to Swindon. As for the length of the sewers under the metropolis, there was no precise answer to be had, beyond 'a lot'. The number of flushers is a less slippery figure. At the time of my visit, it was thirty-nine. Thames Water claims more efficient equipment has reduced manpower needs. The flushers see it differently, muttering about outside contractors doing the job that only they know how to do best, and about asset-stripping in the boardroom. There were personnel cuts after water companies were privatized in 1989, and Thames Water is now onto its third owner in nineteen years. Debates rage still about the wisdom of privatizing companies responsible for providing, in many eyes, an essential public good which costs money to clean and supply.

All the flushers know is that they're heading towards retirement, that the sewer knowledge they carry in their heads is irreplaceable (and unwritten), and that they could use some more staff, though only men like themselves. Sewers have always been a man's world. In London, they're a white, working-class man's world. There are few jobs left that are as monochrome and mono-sexual. There are female engineers who do sewer surveys, sometimes. But no-one can remember a woman applying to be a flusher. Even black-cab drivers – who share the banter, skin colour and accents of the flushers – have reluctantly welcomed some women. But the flushers are not cab drivers, and they've chosen, over the mapped roads above, these mostly unmapped and significantly more dangerous conduits, thoroughfares and bypasses below.

The boundaries of this world are trunk sewers and brick, but they're also the exclusivity of a marginalized occupation. In a scene from *Boys from the Brown Stuff*, a BBC documentary on flusher life, a new flusher tries to chat up a young woman outside a nightclub. He makes the mistake of telling her what his job is. The scene looks set up but her disgust is genuine. 'Does it involve faeces and such? I'm glad I didn't get you to buy me a drink then.'

It's 10 p.m. now. Night is a good time to enter sewers, when businesses – which contribute the biggest volume of waste – have closed. Night is when dangerous sewers are as safe as they can be. This first sewer is safer still, because the flow has been diverted to allow us access. It would only be a metre or so high normally because the Fleet sewer, formed when the filthy Fleet river was enclosed with brick, isn't one of the bigger ones. Some tunnels are several metres in diameter and wide enough to drive a Mini through. Some are barrel-shaped, some shaped like Wild-West wagon canopies. The Fleet is a brick egg. (Elliptical shapes are strong and encourage the flow of water.)

In the Fleet, we are to hunt for leaks. Water systems always leak. In 2006, Thames Water lost 915 million litres of clean water from its drinking-water pipes, an amount that a City of London inquiry thought was 'staggering'. The job tonight is to see whether water is leaking into the sewers from drinking-water pipes nearby. Rob Smith is my guide. He's a tall, powerful-looking man, not far from retirement, who spent twenty years building tunnels before moving into sewers. Both his chosen occupations probably explain his decision to live on the coast while working in London, a commute of a few hours a day. He likes fresh air.

Smith is now a senior engineer, a few rungs on the wastewater ladder above a flusher, and he doesn't need to go into the sewers any more. But he says, 'I can't be responsible for the safety of my men without knowing the environment.' So down he goes, regularly enough, sometimes with a journalist or prince in tow. Thames Water runs open days at its Abbey Mills pumping station where visitors are served sandwiches and tea then led into the trunk sewer below. (It is thought sensible to serve food before sewers, not the other way around.) Smith has seen all sorts. 'Prince Charles came once, down the sewers. We've had Lords and Ladies. They're all the same once they get down there. If anything happens, and someone needs to be pulled out, nobody gets priority. A sewer is a great leveller.'

Smith enters first, nimble and fast down the ladder. I romantically assume he's gone before me because his nose can sense danger. But he has lost much of his sense of smell from hydrogen sulphide exposure. This is annoying above ground but potentially lethal beneath it. Smith's fatigued nose will be backed up by his turtle.

My best defence is a big, long rope that links my harness to a hoist above. A line of life. I'm glad of it, being so weighed down with turtle and tungsten that a stride over to the manhole takes twice the effort. I follow instructions: sit on the pavement. Swing legs over onto the ladder. Grip the manhole cover for purchase. Go down, as slowly as possible. Really, really slowly. 'Take your time!' the flushers shout down, because I am precious cargo. 'No-one gets killed in my sewers,' Smith says. 'Not in, under or above them. It causes a hell of a lot of paperwork.'

The ladder is rusty and damp. The rungs are far apart. I'm apprehensive, waiting to be hit by a stink, but nothing comes. 'That's what people do,' says Smith. 'They get down, take a sniff, say, "Is that poo?" I say yes. They say, "It doesn't smell much does it?" They think that because when they go to the toilet, it smells, that this will too. They think it'll smell like three million toilets.' But this is not a bad smell. It's musty, cloying and damp, but it doesn't stink. It's diluted, after all. Without water, the average human produces 35 kilograms of excrement and 500 litres of urine a year. Add toilet flushes, and the total jumps to 15,000 litres. Thanks to the WC, the flow is ninety-eight per cent water.

Down below, I am unhooked. My safety now depends on the monotonous beeps of the turtle, which signal safe air, and on the men in front and behind me. They set off with the walk of the flusher and I do my best to copy. The sewerman does not walk like an ordinary man. Lifting the feet, as a normal gait requires, risks kicking up the flow and splashing foul water on yourself or your workmate. For this reason, and to get better purchase on slimy brickwork, it's better to glide. Feet close together; buttocks clenched (as tightly as the lips, which are best kept pursed to defend against splashes); smallish steps. It's mincing that manages to be macho. I try to glide satisfactorily while I take in the sights. There are bricks, shadows and light. There is a surprising amount of beauty, which explains why sewers have their obsessive fans, and why they are so beloved of film-makers. What lighting director wouldn't want to rise to the task of shadowing a Harry Lime in black, white and grey menace?

The men have their eyes cast upwards, looking for the incursion of leaked water. Mine look the other way, into the stream. I am nervous about what I might see and curious about what I might recognize. There's a floating, bloated tampon. There goes part of a polystyrene cup. I find myself peering for brown solids, alert and excited, like a kid with a fishing rod. In olden days, sewers had hunters called toshers. They moved into the sewers from the banks of the river, in search of discarded riches. Sometimes they found gold; sometimes they lost their lives. There are still sewer hunters today, and there is cause: the flushers find all sorts in the flow. Bits of motorbikes (easily shoved down a two-feet wide manhole), prams, goldfish. Coins, sometimes, and jewellery. Cell phones by the hundred (one recent survey concluded that 850,000 handsets a year are inadvertently flushed down British toilets). That's all due to haplessness; there's also ignorance. Wastewater utilities have had a long-running 'Bag it and Bin it' campaign to educate people about what they shouldn't flush. The list includes condoms, tampons and applicators, sanitary towels, panty liners and backing strips, facial and cleaning wipes, nappies, incontinence pads, old bandages, razor blades, syringes and needles, colostomy bags, medicine, toilet roll tubes and tights. Bras are also unwanted: in June 2007, a bra and knicker set flushed down a toilet clogged sewers in County Durham, collapsed a road and caused £15,000-worth of repairs. 'Throwaway society,' says Smith. 'My goldfish has died? Throw it down the toilet. My hand grenade doesn't work? Throw it down the toilet.'

Hand grenade? It belongs in Smith's best sewer anecdote, which he has told before and will tell again. He was working with a gang in the mid-level sewer near Greenwich when a flusher handed something to him. It was filth-encrusted but then he made out its shape through the muck. 'I thought, oh shit.' He couldn't see if the grenade was live, but if it had been, it could have blasted a hole up to the sewer above. The gang would either be blown up or drown or both. Smith climbed up the ladder one-handed, having warned the lads above, who disappeared. He lobbed it down an embankment and hoped for the best. 'The next day,' he says, 'a policeman phoned to ask me why I'd done that. I said, I didn't have a choice. I asked him if it had been live, and he said, "You don't want to know," so I presume it was.'

I love sewer anecdotes as much as the men like telling them. They are rich and funny, with a spirit mined from working at extremely close quarters – flushers have to pull and push each other in tight spots, in splendidly intimate isolation – in a job that gets only mockery and disregard from the public. The jokes are revenge. The writer Sukhdev Sandhu met a flusher who 'remembers the night he emerged from a sewer at Leicester Square dripping of filth and shit only to find a young woman tourist peering at him. He held out his hand. "Smell that. That's Canal No. 5, that is."'

Humour helps because the work is hard. The pay isn't great, there are shampoo bills, and then there are the daily grievances like cotton wool buds. 'They are the bane of our lives,' says Smith. 'If someone had searched for something that could clean your ear and also stick perfectly in the 6mm holes of [a filter] sieve, they couldn't have done better.' He shines his light on a pipe mouth to one side, encased with something I can't recognize, dripped solid like stalactites. 'Concrete. Unbelievable. Someone's just poured liquid concrete down a drain.' The liquid has now hardened, embracing and defeating the black pipe it arrived down, a sign of short-sighted selfishness.

The men stop to shine light at roof bricks, searching for cracks. While they look at the bricks with a purpose, I just look at the bricks. Smith is proud of them. 'If you had a garden brick wall,' he says, 'think of the condition it would be in after fifty years. These are over a hundred years old, and they have sewage flowing through them constantly.' He gives them his considered engineer's opinion. They are 'in pretty good nick'.

This sewer is relatively young at a century old. The core of

London's sewer network was built between 1858 and 1866 by a man whose name is now venerated only amongst flushers and historians, though he was probably the greatest of the famed Victorian engineers such as Isambard Kingdom Brunel, bridge-builder, or the locomotive-designing Stephensons. The man who built London's sewers, though, is as obscure as the network he constructed.

Since its beginnings as a trading centre on a useful river, London dealt with its excrement as other settlements did, with what is known today as 'on-site sanitation'. In short, this meant that its citizens generally did their business in a designated, confined place. It was a private matter unregulated by any authority (and done mostly in a privy, from the French word *privé*). Privies were used alongside cesspools and middens (dung heaps). The cesspools were designed to leach their liquids into the soil, leaving the solids to be collected by 'gong-fermors' (a corruption of 'gunge farmers') and sold to farmers as manure. It was a sensible system with much to admire. Nothing was wasted; everything was recycled. The nutrients ingested by humans in food were taken from their cesspits and placed back into land that would grow more food, which would be consumed by more humans, who would in turn produce more useful 'waste'. It was a harmonious recycling loop that also managed to be lucrative. It satisfied the demands of nature and of capitalism. But it did not work perfectly.

The private matter of excretion spilled into public life in many ways. There were unemptied, overflowing cesspools, like the one in which Samuel Pepys trod in 1660, when he ventured into his cellar to find it filled with the contents of his neighbour's privy. There was the common practice of slopping out, when chamber pot contents were flung from windows in the early morning, which made for unpleasant streets, especially since pavements were not common. There is a theory that the popularity of high heels dates from this time, something that might amuse the Yahoo sewer-footwear fetish group, as would the fact that the uppers of Parisian sewer waders were popular with boot-makers. They valued the leather – hardened by contact with fats and acids in sewage – and turned it into ankle-boots for fashionable ladies who remained happily ignorant that their new purchases had spent years wading through the most unfashionable muck.

By modern standards of smell and hygiene, London was disgusting. So was everywhere else. Over the Channel in Paris, contemporary accounts tell of grand aristocrats regularly soiling the corridors at Versailles and the Palais Royal. At Versailles, the garden designer Le Nôtre deliberately planted tall hedges to serve as de facto toilet partitions. The eighteenth-century writer Turneau de la Morandière described the Versailles of Louis XV as 'the receptacle of all of humanity's horrors – the passageways, corridors and courtyards are filled with urine and faecal matter'. Waste matters in the Kremlin were no better, and toilet facilities only improved because it was feared all that excreta would corrode the gold.

Nonetheless, London and Paris both continued on their smelly way until population growth intervened. With industrialization and rural migration, London grew from 959,000 residents in 1801 to 2.3 million in 1851, making it the largest city in the world. The on-site system could no longer cope. There was too much waste to dispose of and inflation didn't help: the cesspool emptying fee was by now a shilling, twice the average labourer's daily wage. Also, the gradual introduction of the flush toilet increased the amount of water to be dealt with. Faced with expense and hassle, people did what people still do, and illegally dumped their cesspool contents into the nearest pond, river or sewer.

London had had sewers for centuries. Henry VIII had issued the first Bill of Sewers in 1531, which gave 'the loving Commons' the powers to appoint Sewer Commissioners, Tudor environmental health inspectors who inspected drains and gutters. But neither the commissioners nor the sewers they protected were concerned with human excreta. The word 'sewer' either derives from 'seaward', according to one source, or, according to the compilers of the Oxford English Dictionary, from the Old Northern French *seuwire*, meaning 'to drain the overflow from a fish pond'. Somehow, in a way obvious only to etymologists, *seuwire* in turn derived from the Latin *ex* [out of] and *aqua* [water]. Sewers have always been a carriage for dirtied water, but the degree and manner of dirt has changed. The modern assumption that sewers carry sewage is relatively new, as is the presumption that waste and water have always gone together.

There were some in antiquity who decided water was a clever way to carry away the contents of their latrines. Primitive forms of the flushing toilet, together with channels to carry foul water away, were found at the 3700-year-old palace of King Minos at Knossos. (This allowed one twentieth-century Englishman to wonder why his Oxford college 'denied him the everyday sanitary conveniences of Minoan Crete'.) The Romans had the Cloaca Maxima, a large city sewer that was cleaned by prisoners of war.

But most ancient societies did not think of using water to transport waste because they didn't need to. The volume of waste and of people could be satisfied with on-site containment and removal services. Even after toilets became popular, it remained illegal for London's citizens to connect their waste pipes to the sewers. It had to go somewhere. By 1840, as the Victorian builder Thomas Cubitt testified before the Parliamentary Select Committee into the Health of Towns, 'the Thames is now made a great cesspool instead of each person having one of his own'.

In these conditions, diseases thrived happily and fruitfully. Faeces carry nasty passengers, and one of the worst is cholera, which arrived from India by ship in 1831. Cholera's primary vehicle is the excrement of humans, who act like inadvertent seeders of the bacteria by expelling diarrhoea violently and relentlessly. In a good sanitary system, where faeces are kept separate from drinking water, cholera would be contained. But in early-nineteenth-century London, when five of the city's nine water companies drew drinking water from the great cesspool of the Thames, cholera was in its element. The first epidemic of 1831 killed 6536 people. In the 1848–9 epidemic, 14,000 died in London, and 50,000 nationwide. Cholera's increased murderous performance was due, ironically, to sanitary reform.

The Victorian century gave us many wondrous things, but one of my favourites is the now-lapsed vocation of 'sanitarian', a word taken by men who occupied themselves with the new discipline of 'public health'. The most famous was Edwin Chadwick, a difficult character who left a legacy of reforms that were magnificent – the 1848 Public Health Act, for one – but also mistaken and deadly. In Chadwick's landmark 1842 *Report on the Sanitary Condition of the Labouring Population of Great Britain*, a Victorian best-seller, he condemned the filth in which working classes were forced to live, its effects on their health and the consequent losses to the economy. (Henry Mayhew, in a letter to the *Morning Chronicle*, wrote of meeting a woman in cholera-ridden Bermondsey who said simply, 'Neither I nor my children know what health is.')

Chadwick decided the solution was to organize and expand the sewer system, but to use it for sewage – a word newly invented – and to discharge the sewage into the Thames. It might hurt the river, he reasoned, but it would save people's health. Sewers were built and did as he said they would. And the Thames ran browner and thicker, and people drank it, and cholera loved it. There were fulminations against filth in newspapers and parliament, but nothing was done. The medical establishment, in these pre-Pasteur times, was still convinced that disease was spread by contagion via miasmas, or bad air.

It took a long dry summer to force change, and because of the foulness of the air, not of the water. In 1858, the weather and the sewage-filled Thames came together disastrously to form the Great Stink, when the river reeked so awfully, the drapes on the waterfront windows of the Houses of Parliament were doused with chloride to mask the smell. Politicians debated with their noses covered by handkerchiefs. After prevaricating for years, parliamentarians debated for only ten days before signing into law the Metropolis Local Management Act, which set up a Metropolitan Board of Works to sort out the 'Main Drainage of the Metropolis'.

Joseph Bazalgette was the Board's Chief Engineer. He was a small man with excessive energy. His plan was grand; enormous main sewers would run parallel to the river on upper, middle and lower levels. They would be fed by a vast network of smaller sewers, and the whole flow would be conducted by gravity and sometimes by pumps (London is partly low-lying) to two discharge points, at Barking and Crossness, in London's eastern reaches. There, the city's sewage would continue to be dumped into the river, but suitably far from human habitation. Dilution, as the engineer's mantra still goes, would take care of pollution. Construction lasted nearly twenty years. By then, Bazalgette had used 318 million bricks, driven the price of bricks up fifty per cent and spent £4 million, an enormous sum (£6 billion in modern money). He'd also built the Victoria Embankment along the way, reclaiming land from the river near Westminster and running a sewer through it. For all this, as Stephen Halliday writes in *The Great Stink of London*, he should be considered the greatest sanitarian of all. His sewers may have saved more lives than any other public works. But his efforts have been rewarded with a small plaque on his embankment, a mural in some nearby public toilets and two streets in the far-off London suburb of New Malden. There is no statue or public thoroughfare celebrating his Main Drainage of the Metropolis, though Bazalgette arguably did more than Brunel to shape modern life.

*

The flushers love Bazalgette, particularly because he built his sewer network with twenty-five per cent extra capacity to allow for population growth. But they have more pressing thoughts than their hero's cultural legacy. However ingenious Bazalgette's design was, a system built for 3 million must now cope with the excreta and effluent of 13 million people and hundreds of thousands of industries. Bazalgette couldn't imagine there would be so many houses, with so many toilets using so much water. This much that can be thrown away, when it needn't be. He certainly didn't account for sewers that take everything away being defeated by takeaways of another sort.

There is a mid-level sewer that requires inspection. A nearby park is genteel in darkness, and there is beauty down the manhole too, in the form of a spiral brick staircase, glistening with damp and other things best not inquired into. 'This is an original Bazalgette,' says one of the two men preceding me. Then he stops dead.

'Fat.'

Fat?

'Here, look.'

The stairs are stuffed with blocks of solid, congealed fat. The industry term is FOG, for Fat, Oil and Grease. Flushers hate it even more than cotton wool buds. Faced with this degree of FOG there is only defeat and retreat. Up above, Dave the flusher lets rip. 'Fat! It costs millions to clean up. Restaurants pour it down the drains, it solidifies and it blocks the sewers.' They used to use road drills to remove it, he says – 'big RD9 jobs!' – until new Health and Safety regulations came into force, and jobs that had been done for years were judged now to be too dangerous. Flushers still talk of the Leicester Square fat blockage that took three months to remove. Once, Dave's gang was hammering away at a whole wall of FOG, and another gang was doing the same at the other end, until the wall started shifting and nearly squashed the gang on the other side.

Flushers are phlegmatic about faeces or toilet paper or condoms. But they hate fat. 'That's what smells,' says Dave. 'Not shit. Fat gets into your pores. You get out and you have a shower at the depot and you smell fine, then you get home and you smell again.' They grimace. 'Disgusting stuff.' It is also expensive stuff. Half of the 100,000 blockages every year in London are caused by it. It costs at least £6 million a year to remove. 'Contractors do it now,' says a flusher, before muttering, 'or they don't, more like.' High-pressure hoses flush out some blockages. Thames Water has been trying out robot fat-removers and already uses remotely operated cameras to see what's what, but for now the best weapons against an unceasing and superior enemy are water, force and curses. Prevention would be better. Restaurants are supposed to have fat traps, but enforcement is minimal. It costs money to get collected fat carted away, so many restaurants dispose of it down the sewer instead. Leicester Square's restaurants are no more responsible than Victorian London's cesspool owners were. Who's going to find out? Most sewers are only visited when something goes wrong, and monitoring is light. Sewer workers are fire-fighters: they respond to crisis. In most areas of the UK, only twenty per cent of sewers are inspected regularly, and by the end of this century, many of the UK's 186,000 miles of sewers will be 250 years old. They may be in pretty good nick, but sometimes they don't work.

In an average year in the UK, 6000 homeowners find sewage has backed up into their houses or gardens. Consider for example the troubles of Sonia Young, who spent 100 days in total cleaning her garden of its unintended sewage pond feature, or new mum Elizabeth Powell in Bath, forced to escape upstairs with her two-week-old baby from a flood of sewage that reached her knees. In 2003, the Court of Appeal at the House of Lords, the UK's highest-level judiciary body, heard the case of Peter Marcic, resident of Old Church Lane in the London suburb of Stanmore. Between 1993 and 1996, Marcic found sewage backed up in his garden once a year. It happened again: twice in 1997, not once in 1998, four times in 1999 and five times in 2000. During the hearings, the Lords seemed shocked by several things: that a modern-day wastewater treatment infrastructure can still spew shit into a residential home, and that it is considered normal. That 'sewerage undertakers', as the water utilities are known, are only obliged to compensate the homeowner for the cost of their annual sewerage rates, usually around £125. That insurance companies, faced with costs of between £15,000 and £30,000 per sewer-flooding incident, sometimes refuse to pay up. Under the 1875 Public Health Act, still in force, local authorities are obliged to make 'such sewers as may be necessary for effectually draining their district'. 'Effectually' is vague enough to leave room for loopholes, and to get out of infrastructural upgrades. In some ways, this is understandable, as water utilities get no extra public subsidy for infrastructure costs and must pay for them out of water and sewer rates. But any rise in bills unfailingly causes public outrage. (A few months after raising water bills by 21 per cent in 2005, Thames Water's four directors were awarded bonuses totalling £1.26 million, a rise of 62 per cent on the previous year.)

A 2004 parliamentary committee was appalled by testimony from water industry officials about sewage backups. 'Would you say,' enquired a committee member of the head of England and Wales' water regulators Ofwat, 'that sewage ending up in your living room is about the worst service failure that can happen to anybody?' The man from Ofwat had to agree. 'Short of threats to life and limb and health,' he admitted, 'it is one of the most unpleasant events that can happen to any household.'

Bazalgette's sewers may have saved London from cholera, and made miracles out of brick and water. But even he couldn't defeat decay, pinched resources and a failure to upgrade. 'If Bazalgette hadn't built his sewers when he did,' Rob Smith tells me, 'we would –

literally – be in the shit today.' If Bazalgette's sewers aren't maintained, we will be again.

It's a hot afternoon in Queens, New York, and for the first and probably last time in my life, I am stopping traffic, with the assistance of half a dozen fit young men and four large trucks belonging to the New York City Department of Environmental Protection (DEP). The men have been asked to show me a regulator sewer, a tour that took months of begging to arrange. The process began with a redoubtable woman in the DEP press office who declared that there was totally no way I'd ever get into the city's sewers, and did I know how many people phoned her to ask the same question? But the redoubtable woman could be bypassed, and the bypass ended up in the office of Deputy Commissioner Douglas Greeley, an affable man with a hell of a job, because he is in charge of New York City's fourteen wastewater treatment plants (London, by contrast, has three).

When I'd called him from London to arrange the appointment, he'd been helpful but doubtful. 'You'll have to do several hours of close-confinement training. Then you'll need to get a security check.' Then he said '9/11!' as if he didn't need to say more. Precaution is understandable and probably overdue: terrorist attacks on drinking water supplies are usually planned for, but not on sewage facilities. (When an employee at a Washington, DC treatment plant showed me the railway trucks that until recently took liquid chloride down to the river to purify the effluent, he said, 'We really dodged a bullet there. Any terrorist could have blown up those trucks and killed 10,000 people. Luckily, terrorists are stupid.') Sewers have always had security issues: Leon Trotsky ordered Moscow's sewers to be checked for opponents with bad intentions. The most sensitive sewers in London – which Rob Smith wouldn't identify on a map, but which definitely include one running under Buckingham Palace – have sensors linked to police stations, so that flushers doing a job below who haven't alerted the right control centre risk emerging to find several gun barrels pointing in their upcoming direction.

Eventually, vetting was deemed done, my time spent in London's sewers counted as close-confinement training, and Greeley said he could pass me on to 'a happy Irishman who will look after you'. The happy Irishman is in charge of Collections North, a department with a dull name and an important job, because they keep the pumping stations and tide gates working. Pumping stations pump up the flow when gravity isn't enough, or when gravity is going in the wrong direction. Some are three storeys deep. The tide gates date from the days when sewers – wooden pipes back then – ran under the piers where ocean liners docked. Tide-gate discharges – 'outfalls' – were accepted for decades, Greeley tells me, 'until passengers thought the smell was too great when they were getting off the boat. People would go to the beach and there would be something like black mayonnaise all over it and it was like a horror show.'

Greeley has shelves full of water and sanitation books in his office, a piece of original wooden water pipe mounted on his wall, and a lively interest in New York water history, clean and foul. He can and does talk about it for hours. He explains that sewer construction was slower than London's and more piecemeal. In the nineteenth century, each of the five New York boroughs had autonomy and a president. Each president got around to sewer construction when he felt like it. It wasn't considered urgent. There was no Great Stink, no cholera to focus priorities. Drinking water was a different matter. 'They were thinking, damn the economics of it, we're going to build for a hundred and something years from now. That thinking went into our older structures, the reservoirs, the aqueducts. The water system was built for the ages. The sewer system on the other hand? "Only do what we have to."'

Manhattan's and Brooklyn's sewers were rationally laid out thanks to a sewerage commission that travelled to Europe to learn from

Hamburg (the first European city to lay modern sewers) and London. Queens, Staten Island and the Bronx fared worse. 'They used to be farms. By the time they opened up, the city couldn't catch up. It was always anticipated that the city would be able to catch up, but that was sixty years ago.' Greeley says his sewers are also in quite good nick, as they are regularly sprayed with concrete, which helps prevent wear and tear. That's not to say that the DEP couldn't do with more money for upgrades. The American Society of Civil Engineers grades the nation's infrastructure every few years. In 2000, wastewater infrastructure got a D. By 2005, it was a D minus. In 2000, the United States Environmental Protection Agency (US EPA) estimated that a quarter of the nation's sewer pipes were in a poor or very poor condition. By 2020, the proportion of crumbling, dangerous sewer pipes will be fifty per cent.

This isn't the only pressing problem. Greeley's life is made more difficult because when the nineteenth-century sewerage commission came back from Europe and made its decision, it was the wrong one. At the time, there were two major design choices for sewer systems. The first separates sewage from storm-water and is called a Separate Sewer System (SSS). The second does not. A Combined Sewer System (CSS) puts water from all sources – street, toilet and anywhere else – into the same pipes. It is cheaper and easier to construct, which is why New York's sewer designers probably chose it. But it has one powerfully weak spot. Rain.

Sewer designers try to plan for excessive rainfall by installing storm tanks at points along the system and as emergency reservoirs at wastewater treatment plants. When more rain than expected falls, it can be held safely and the sewers will not flood. But a tenth of an inch of rain, falling in a short space of time, can overwhelm the tanks. Then, the system does what it's designed to do in such circumstances: it discharges raw, untreated sewage into the nearest body of water. Such discharges are called CSOs (Combined Sewage

Overflows) and they are far more common than most people think. In New York, according to the environmental group Riverkeeper, there is generally one CSO a week, and the average weekly polluted discharge is about 500 million gallons, an amount that would fill 2175 Olympic-sized swimming pools. Nationwide, according to the EPA, the wastewater industry discharges 1.46 trillion gallons – I can't conceive how many swimming pools that is – into the country's waterways and oceans.

'Look,' says Kevin Buckley. 'It's either discharge or it comes up in people's basements.' Buckley, the happy Irishman who has organized the traffic-stopping exercise in Queens, has taken me over the road to see the nearby outfall into Jamaica Bay. We watch a crab tootling between the booms that are supposed to direct wet weather discharge into the bay, next to a sign that tells people to call 311, New York's non-emergency hotline, if they see sewage pouring out in dry weather. 'That's a no-no,' says Buckley. But wet weather discharge is normal. It's how the system works, whether people know it or not. Sewer designers calculate their system capacity to cope with storms and floods. New York's sewers, built in drier, less globally warmed times, were built to cope with a maximum of 1.75 inches of rain falling in an hour. Times and the weather have changed. Buckley only has anecdotes to back him up, but he swears storms are getting more frequent and more intense.

On August 8, 2007, 3.5 inches of rain fell in 2 hours in Manhattan, and 4.26 inches in Brooklyn. The subway system failed: this was more water than their pumps could cope with, and the tracks were flooded. The Metropolitan Transport Authority blamed the DEP, saying it couldn't discharge the water because the sewers were already full; the DEP blamed the MTA. In fact, as the then-Governor Eliot Spitzer said, neither was really to blame, because 'we have a design issue that we need to think about'. In its report *Swimming in Sewage*, the Natural Resources Defense Council expressed exasperation that 'the nation at the forefront of the information age has about as clear a view of the quantity of sewage that leaks, spills and backs up each year as we do of the sewage pipes buried beneath our feet'. When a catastrophic overflow happened in London in 2004, and 600,000 tonnes of raw sewage poured into the Thames, people did notice. Fish died in their hundreds. Newspapers called it the Lesser Stink. The newly formed Rowers Against Thames Sewage (RATS) organized a rowing event on the same stretch of river that hosts the Oxford-Cambridge Boat Race. The Turd Race saw two boats – Gashaz and Biohaz – tow giant inflatable faeces for 800 metres, with the rowers wearing gas masks. Biohaz stormed to victory. A parliamentary inquiry expressed 'abhorrence at this legitimized pollution and the depressing attitude with which it is accepted', and eventually, after fifteen years of procrastination, the government approved plans for a £2 billion interceptor stormwater tunnel to run under the Thames.

And what about New York, city of confident skyscrapers built over an increasingly fragile infrastructure? The subways started working again, a New Yorker friend tells me, 'and that was it. Everyone forgot about it.'

Back in Queens, the men are ready to go down the hole. I'm wearing a Tyvek suit, made from the same material that weather-proofs houses under construction. I don't have breathing equipment, because this is a regulator chamber – a sort of sewer intersection – with a viewing platform, and we aren't going deep. Anyway, when I asked for a turtle, I got strange looks. (Later, I discovered that 'turtle' is American sewer-worker vernacular for faeces.) No helmet is offered, because the chamber doesn't warrant one, though the roaches might. I don't mind rats, but I hate roaches. Down the ladder, the team leader, a handsome ponytailed man named Steve, shines his torch up at the corner, where several dozen of the biggest roaches I've ever seen immediately set about scurrying into safe darkness. Steve grins. 'It's OK, they're not roaches. They're waterbugs.'

What are waterbugs?

'Roaches on steroids.'

The day before, Steve had entered a sewer he'd never been into before – not unusual, when there are 6000 miles of network – and the walls were moving. 'You shine your light and they move, but if you leave them in peace, they'll leave you alone too.' (He always tucks his ponytail into his shirt collar in case.) The same respect goes for rats, in the main. 'You're going into their home, so you treat it with respect.' Precaution doesn't mean indulgence, not if they're even half the size that flushers say they are, or if they're anything like the rats described 160 years ago to Henry Mayhew by a man from a Bermondsey granary: 'Great black fellows as would frighten a lady into asterisks to see of a sudden.'

I'd seen one rat in London's sewers, and no asterisks were provoked. The flushers must have been disappointed, because they started on the rat tales as soon as I got out of the hole. There was the story of fearsome Jack, a flusher famed throughout London for his ability to kill with his hard hat. Keith preferred his shovel. Dave had had one run up his arm on a ladder. Happy had seen one the length of his forearm. Honest.

The New York collections men are no different. They see rats all the time, and despite professing respect for their habitat, Kevin will often dispatch them into the flow with a whack from a bat he carries. 'They can swim, but it's so fast, they won't survive that.' The worst thing about rats, says Steve, 'is waiting for that big wet slap on your back'. 'No,' says Kevin. 'It's knowing you're being watched, but not knowing who's watching and from where.' London's sewer rats generally run away from humans. New York's don't. 'They come at you,' says Steve. I must look disbelieving, wondering if flusher-men and fishermen exaggerate alike, because the men are indignant, and look to each other for confirmation. 'Really! They'll jump on you, no problem.' Kevin swears there's a rat near the river who's so fearsome, it once climbed up the manhole ladder. 'And the rungs are very far apart.'

The sewers also produce less troublesome fauna. In a tank at New York's Ward's Island treatment plant, twelve turtles live happily in clear water, having been rescued from the grit chambers that screen the flow before it heads under the East River to a facility in Brooklyn. A Russian worker saunters past and mutters 'good soup', but the turtles are well looked-after, especially considering that soup is what they would have become if they'd gone through the gritters. The turtles are the small, pet-shaped variety. Huge snapper turtles end up in the system too, but they're taken and put back in the river. Or so they say. They would also make good soup.

New York's sewer workers are a cheery lot. Morale seems healthy, and better than that of their London colleagues, who told me gloomily that they didn't like coming to work any more, and that 'shit [was] going to pot'. Steve's wife doesn't like 'the shit factor', and refused to watch a TV programme on dirty jobs that would have showed her what he did, but he seems unbowed. He likes his career and he thinks it's a valuable one. He tells me that of course he grew up dreaming of becoming a sewage treatment worker, before his sarcasm is leavened with a smile.

He could have, if he'd watched enough TV re-runs. Unlike their British counterparts, American sewer workers can reflect in the glory of a much-loved sitcom character from the 1950s. Millions of Americans remember – and loved – the character of Ed Norton in *The Honeymooners*, a sewer worker with an endless supply of wastewater witticisms, most of them involving lying back and floating. Sewer-worker pride is also fed by the Operators' Challenge, a nationwide annual competition set up by the Water Environment Federation, an industry body. Wastewater workers compete in several events, such as rescuing from a sewer a mannequin in danger; fixing machinery; and answering technical questions in Wastewater Jeopardy, a version of the US quiz where contestants are given answers for which they have to guess the question. (Question: The minimum design velocity in sewers to prevent solids from settling in the collections system. Answer: What is two feet per second? Question: The mixture of micro-organisms and treated wastewater. Answer: What is mixed liquor?)

The competition is taken seriously – there are Operators Challenge trophy cabinets in every treatment plant I visit – even if the team names lack gravitas. The Ward's Island Ninja Turtles compete with the Bowery Bay Bowl Busters and the Tallman Island Turd Surfers. The media treat it with humour, referring to the Sludge Olympics, and coverage brings prestige. 'It's genuinely good for improving skills,' says Buckley, who adds that much of the work is achieved 'with brute strength and ingenuity'. He tells with pride of his most ingenious hour, when he traced a catastrophic spill of boiler oil in a local creek two miles back up the sewer line, right to the basement of the apartment building that was sending it into the sewer. He got commendations; the offender got a $3 million clean-up bill. He says his investigative technique involved sticking his head down manholes and stopping at the first clean spot of sewer he saw.

This is a skilled job, and it's sought after, though not for the salaries. The newest team member is Edwin, a young tattooed man whose low rank is obvious because it's his leg that men grab for balance when they're going down the holes. Edwin earns $15 an hour. The most senior crew member only gets $21. I suspect they'd earn more cleaning toilets. More attractive are stability and benefits, crucial in a country where the only healthcare on offer has to be paid for. That's what attracted Buckley, when he got off the plane from

London in the 1970s, and didn't want to be 'the stereotypical Irish navvy'. The only thing lacking for job satisfaction is a proper New York nickname.

At a manhole in La Guardia airport, Buckley is showing me another tide gate, shining light into the hole with a mirror and sunlight ('better than any flashlight') when a Port Authority cop stops by. He asks what we're doing and when Buckley replies, 'Looking for alligators,' nods with no apparent disbelief before moving in for a peer. I ask the cop why they're known as New York's finest, why fire-fighters are New York's bravest and even prison officers at Riker's Island are New York's boldest, but the men who keep sewage flowing, and keep disease away, have nothing. He shrugs. He doesn't know or care. Buckley laughs. 'We're New York's stinkiest.' Sometimes New York's bravest can't do without New York's stinkiest: Douglas Greeley remembers the police asking for his men's help in retrieving a dead Mafioso who had been thrown down a manhole. Another time, the item being retrieved was a broomstick discarded by certain police officers who had used it to sodomize a Haitian immigrant named Abner Louima. 'It was very humid, and the police department internal affairs division had spread canvas sheets out on the street. They closed the street and we scooped every catch basin, and we were pulling out all kinds of broomsticks. We had to lay them down on the canvas and then they would categorize them, measure them and do samples. In 95 degree weather.' They found it. Louima was eventually awarded $5.3 million in damages against the city, the largest police brutality settlement in its history. The contribution of sewer-workers to the investigation went unnoticed.

They could be New York's damnedest, working with a system that is heinously expensive to maintain and upgrade, excessively wasteful of water and easily defeated by less than half an inch of rain. Greeley knows that flooding could be minimized if the rain had somewhere else to go beside a sewer, such as into the earth. But New York – and London – have blocked off all natural drainage by concreting over much of their surface area. Patio gardens also have a lot to answer for. Greeley talks wistfully of Seattle's Street Edge Alternative (SEA) streets, where asphalt is removed and replaced by wide borders of whatever encourages the natural percolation of water downwards (earth, turf, pebbles). Similar plans have been proposed as part of New York Mayor Michael Bloomberg's Sustainable City initiative. Even in London, where homeowners have been allowed to pave over the equivalent of twenty-two Hyde Parks in ten years, the government has announced that covering earth with anything other than porous materials will now require planning permission. I ask Greeley if he has the money for things like SEA streets. 'No. That's the tragedy.'

I also ask him the question I put to everyone I meet who works in or with wastewater. If they had to design the system again from scratch, would they do it differently? Would they, as former President Teddy Roosevelt did, question the original concept of flushing? 'Civilized people,' Roosevelt said in 1910, 'ought to know how to dispose of sewage in some other way than putting it into the drinking water.' Greeley considers the question with a long pause. 'It's true that waterborne sewerage is very problematic.' But a return to on-site sanitation, whether privies or private treatment plants, is no solution. 'People wouldn't look after them properly. There'd be disease outbreaks.'

Before I leave, after Greeley has loaded me down with lapel pins that are miniature New York Sewer manhole covers, he plays me the 'Song of the Sewer', sung by Art Carney from *The Honeymooners*. We listen to it, a rare example of a positive spin on sewer work – 'Together we stand/shovel in hand/to keep things rolling along' – and he ponders his chosen career. The great American sanitary engineer George E. Waring may have written about 'the branch of the Art of

Drainage which removes faecal and other refuse from towns', but who, these days, thinks waste disposal an artistic endeavour? 'A city father,' says Greeley, 'would never say, welcome to our sewer system, isn't it special, we're proud of it. The best I can hope for is indifference.'

World Toilet Expo, Bangkok, 2006 – *Author*

2. THE ROBO-TOILET REVOLUTION

The actress and the gorilla

The flush toilet is a curious object. It is the default method of excreta disposal in most of the industrialized, technologically advanced world. It was invented either 500 or 2000 years ago, depending on opinion. Yet in its essential workings, this everyday, banal object hasn't changed much since Sir John Harington, godson of Queen Elizabeth I, thought his godmother might like something that flushed away her excreta and devised the Ajax, a play on the Elizabethan word 'jakes', meaning privy.

The greatest improvements to date were made in England in the later years of the eighteenth century and the early years of the next, by the trio of Alexander Cumming (who invented a valve mechanism), Joseph Bramah (a Yorkshireman who improved on Cumming's valve and made the best lavatories to be had for the next century) and Thomas Crapper (another Yorkshireman who did not invent the toilet but improved its parts). In engineering terms, the best invention was the siphonic flush, which pulls the water out of the bowl and into the pipe. For the user, the S-bend was the godsend, because the water that rested in the bend created a seal that prevented odour from emerging from the pipe. At the height of Victorian invention, when toilets were their most ornate and decorated with the prettiest pottery, patents for siphonic flushes, for example, were being requested at the rate of two dozen or so a year.

Nonetheless, the modern toilet would still be recognizable to Joseph Bramah. He could probably fix it. Other contemporary inventions like the telephone have gone through profound changes (it's difficult to think of Alexander Graham Bell getting to grips with an iPhone). They have been improved through generations of innovation. The toilet, by contrast, remains adequate and nothing more, though readers of *Focus* magazine once voted it the best invention in history (over fire and the wheel). Compared with other items that are considered necessities – car, telephone, television – the toilet is rarely upgraded voluntarily. Marketers call it a 'distress purchase' because it is only replaced when necessary.

One country treats the toilet differently. Here the toilet is modified, improved upon, innovated. It is a design object, a must-have, a desirable product. Enormous sums are spent on improving its smallest parts. Only here is the toilet given the respect accorded other great inventions.

Three scenes:

On my first morning in Tokyo, I go to get my hair cut. I am the first customer in the shop and talk to the receptionist while I wait. I tell him I'm writing a book about toilets.

'Why?'

I say Japan's toilets are like no other.

'Are they?'

[He thinks]

'Westerners don't like them.'

[He makes a gesture of spray going upwards]
'They don't understand.'

In a tiny bar in Tokyo's Golden Gai district, across the alleyway from Quentin Tarantino's favourite bar, I'm having a conversation with the owner, a hefty, cheery girl from Hiroshima. She has asked what I'm doing here, and I have answered. Oh! That's so interesting! Within five minutes, the entire bar – it holds seven bar-stools, and discretion is pointless – is discussing with great vigour the merits of Japan's two leading toilet brands. TOTO washes better. Yes, but Inax dries better. It's all a question of positioning. My companion, a genteel young woman who runs an art gallery, is amused. 'They are taking it totally seriously,' she says. 'They are genuinely trying to help you. It's nice.'

It is very cold in Kyoto. I have come to Japan in December, in between trips to Bangkok and India, where December is hot. I have not brought enough winter clothing and I am feeling the cold. In Kyoto I walk the streets for a while, dipping into shops for warmth. Eventually it gets to be too much. There's only one option left. Though I have boycotted McDonald's for years, this is where I go because I know they have heated toilet seats. I know they have TOTO.

Japan makes the most advanced, remarkable toilets in the world. Japanese toilets can, variously, check your blood pressure, play you music, wash and dry your anus and 'front parts' by means of an in-toilet nozzle that sprays water and warm air, suck smelly ions from the air, switch on a light for you as you stumble into the bathroom at night, put the seat lid down for you (a function known as the 'marriage-saver') and flush away your excreta without requiring anything as old-fashioned as a tank. These devices are known as high-function toilets, but even the lowliest high-function toilet will have as standard an in-built bidet system, a heated seat and some form of nifty control panel.

Consequently, first-time travellers to Japan have for years told a similar tale. Between being befuddled by used underwear-vending machines and unidentifiable sushi, they will have an encounter that proceeds like this: foreigner goes to toilet and finds a receptacle with a hi-tech control panel containing many buttons with peculiar symbols on them, and a strange nozzle in the bowl. Foreigner doesn't speak Japanese and doesn't understand the symbols, or the English translations that are sometimes provided. Does that button release a mechanical tampon grab or a flush? What, please, is 'a front bottom'? Foreigner finishes business, looks in vain for a conventional flush handle, and then – also in vain – for which button controls the flush. Foreigner presses a button, gets sprayed with water by the nozzle instead and is soaked.

This is the Washlet experience. The Washlet, originally a brand name for a toilet seat with bidet function, has become for the Japanese a generic word for a high-function toilet (though usually translated as Washeretto). In modern Japan, the Washlet is as loved and taken for granted as the Hoover. Since 1980, TOTO, Japan's biggest and oldest toilet manufacturer, has sold 20 million Washlets to a nation of 160 million people. According to census figures, more Japanese households now have a Washlet than a computer. They are so standard, some Japanese schoolchildren refuse to use anything else.

It is easy, for anyone who has not used a Washlet, to dismiss it as yet another product of Japanese eccentricity. Robo-toilets. Gadgetry and gimmickry, bells and whistles. Such sniping ignores the fact that the Japanese make toilets that are beautifully engineered, and that the stunning success of the high-function toilet holds lessons for anyone – from public health officials to marketing experts – whose work involves understanding and changing human behaviour and decision-making. It is instructive because only sixty years ago, Japan was a nation of pit latrines. People defecated by squatting. They did not use water to cleanse themselves, but paper or stone or sticks. They did not know what a bidet was, nor did they care. Today, only three per cent of toilets produced in Japan are squat types. The Japanese sit, use water and expect a heated seat as a matter of course. In less than a century, the Japanese toilet industry has achieved the equivalent of persuading a country that drove on the left in horse-drawn carriages to move to the right and, by the way, to drive a Ferrari instead. Two things interest me about the Japanese toilet revolution: that it happened, and that it has strikingly failed to spread.

TOTO – the name comes from a contraction of the Japanese words for Asian Porcelain – ranks among the world's top three biggest plumbing manufacturers. In 2006, its net sales were $4.2 billion. It has 20,000 employees, two-thirds of Japan's bathroom market, 7 factories in Japan and a presence in 16 countries. With the Washlet, TOTO has given the Japanese language a new word, and the Japanese people a new way of going to the toilet. It is a phenomenon.

I arrange to visit the TOTO Technical Centre in Tokyo. It is a low, sleek building, oddly located in a residential street in an ordinary eastern suburb which has a mom-and-pop hardware shop on the main street, no neon and no visible foreigners. The Technical Centre is described as '[a place] where architects come to get ideas about designs'. It is a show-and-copy emporium, big, spotless and empty of people or architects. Sample bathroom sets gleam in the distance; a row of toilets automatically lift their lids as I walk past, in a ceramic greeting ceremony. Photographs are forbidden, leading me to wonder what an architect who's no good at sketching is supposed to do. But the toilet industry in Japan is a highly competitive business, and the top three – TOTO, Inax and Matsushita – keep their secrets close. My requests to visit TOTO's product development laboratories were politely refused.

My guide is a young woman called Asuka. She works in TOTO's investor relations department and has probably been instructed to deal with me because she went to school in the US for a few years and speaks near-perfect Valley Girl. Perhaps I've met too many engineers, but she doesn't seem like someone who would work in the toilet industry. When she sees a World Toilet Organization sticker on my glasses case, she says, 'On Gucci!' with genuine distaste. She later confesses that, actually, she'd rather be marketing cosmetics. She says that TOTO is a good employer, though I'm disappointed to discover that rumours of certain employee perks are unfounded. They do not get free toilets.

It's Asuka's first time presenting a PowerPoint introduction to TOTO, and despite the occasional sorority phrasing – 'the Washlet is, like, a must-have' – she conveys the facts and figures well enough. The world's biggest toilet manufacturer was founded in 1917, when a man called Kazuchika Okura, then working for a ceramics company, thought it might be a good idea to manufacture toilet bowls. It was not the most obvious business plan. As Asuka puts it, 'Back then, the sanitation environment was terrible here in Japan. We only had wooden toilet bowls.' In truth, they didn't have toilet bowls at all, because squatting toilets didn't have any. Nonetheless, according to the official TOTO history – as told in a comic strip that Asuka gives me, this being *manga*-mad Japan – Mr Okura expressed his desire, in somewhat stilted English, to 'research how to mass-produce sanitary-ware, which are large ceramic items'.

Progress in selling large ceramic items was slow at first. Then came the Second World War, which left Japan with a damaged infrastructure and a determination from planners to build superior housing connected to sewers. This wasn't a new concept: the Osaka Sewerage Science Museum shows a diorama display featuring Lord Hideoshi, a shogun who installed a sewer at Osaka castle 400 years ago. With little thought for chronology, Lord Hideoshi is joined in the diorama by a bowler-hatted Scotsman called William Barton – voiced by an American who learned Scottish from *Star Trek* – who worked in Tokyo University's engineering department and introduced Japan to waterborne sewerage. Still, by the end of the Second World War, only a tiny proportion of the country was sewered.

American forces stationed in Japan, accustomed to flush toilets at home, pushed for the same to be installed in the nation they were occupying. TOTO's toilet bowls sold increasingly over the next forty years, and by 1977, more Japanese were sitting than squatting. This cultural change was not without difficulties. The writer Yoko Mure, in a contribution to *Toilet Ho!*, a collection of essays about Japanese toilet culture (whose title in Japanese apparently expresses the extreme relief of someone who has been desperate for a toilet and finally finds one), wonders 'how the people could use a Western-style toilet. The Western style is the same as sitting on a chair. I had a terror that if I got used to it, I might excrete whenever I was sitting on a chair anywhere, even at a lesson or at mealtimes.'

The new ceramic sitting toilet had other disadvantages. Visiting an outhouse during Japan's freezing winters can never have been pleasant, but at least with a squat pan there was no contact between skin and cold material. The new style changed that: now, flesh had to sit on icy ceramic for several months of the year, a situation worsened by a national resistance to central heating that persists today. A home-grown solution was devised by sliding socks on the seat, but this technique only worked on old horseshoe-shaped seats, which were becoming less common.

TOTO spotted a flawed design that could use some innovation. In 1964, the Wash Air Seat arrived in Japan. Produced by the American Bidet Company, this detachable seat featured a nozzle that sprayed warm water and also blew hot air for drying purposes. In the US, the Wash Air Seat had been aimed at patients who had difficulty using toilet paper or reaching round to wipe themselves. It was a niche item that TOTO thought had mass appeal. But their version failed. It was too expensive. The bidet function was too foreign. History and habit were both against it.

First, there was the bidet issue. In toilet use, the world divides, roughly speaking, into wet (flush) or dry (no flush). In anal cleansing terms, it's paper or water, and, as with driving habits, cultures rarely switch. India and Pakistan have a water culture, so that no visit to the toilet is possible without a *lota* (small jug or cup) of water to cleanse with after defecation. Alexander Kira writes that nineteenth-century Hindus refused to believe Europeans cleaned themselves with paper 'and thought the story a vicious libel'.

In their toilet habits, the Japanese were a paper and stick culture. Wipers, not washers. But they were also a cleansing culture with strict bathing rituals and firm ideas about hygiene and propriety. Keeping clean and unpolluted is one of the four affirmations of Shintoism. Stepping unwashed into a bath, as Westerners do, is unthinkable to the Japanese, where a tradition of bathing communally in cedar-wood baths functions on the assumption that everyone in the bath is already clean.

These hygiene rules stopped at the outhouse door. The Japanese were as content as the rest of the paper-world to walk around with uncleaned backsides. Using paper to cleanse the anus makes as much sense, hygienically, as rubbing your body with dry tissue and imagining it removes dirt. Islamic scholars have known for centuries that paper won't achieve the scrupulous hygiene required of Muslims. In a World Health Organization publication that attempts to teach health education through religious example, Professor Abdul Fattah Al-Sheikh quotes the Prophet's wife, Aisha. She had 'never seen the Prophet [...] coming out after evacuating his bowels without having cleaned himself with water.'

Paper cultures are in fact using the least efficient cleansing medium to clean the dirtiest part of their body. This point was memorably demonstrated by the valiant efforts of a Dr J.A. Cameron, who in 1964 surveyed the underpants of 940 men of Oxfordshire. He found faecal contamination in nearly all of them that ranged from 'wasp-coloured' stains to 'frank massive faeces'. Dr Cameron, though a medical man, could not contain his dismay that 'a high proportion of the population are prepared to cry aloud about footling matters of uncleanliness such as a tomato sauce stain on a restaurant tablecloth, whilst they luxuriate on a plush seat in their faecally stained pants'.

Also, the Japanese didn't know they wanted better toilets. The writer Jun'ichiro Tanizaki reminisced about visiting a privy perched over a river, so that 'the solids discharged from my rectum went tumbling through several tens of feet of void, grazing the wings of butterflies and the heads of passers-by'. But the reality of the Japanese privy had little to do with butterflies. Instead, the average Japanese toilet was known as the four Ks. It was *kiken* (dangerous), *kitanai* (dirty), *kurai* (dark) and *kasai* (stinky). Consequently, it was neither talked about nor acknowledged. This desire for concealing anything to do with defecatory practice surfaces in the common proverb *Kusaimono ni futa wo suru* (to keep a lid on stinky things); in the existence of Etiquette, a pill that claims to reduce odorous compounds present in excreta and is marketed at 'people minding excrement smell'; and in the even greater success of a TOTO product called Otohime (Flush Princess), a box that plays fake flushing sounds to disguise the noise of bodily functions, and is now found in most women's public toilets.

Japan has always had a strong tradition of scatological humour, but it operated beneath polite society levels. These days, times have changed enough for a golden faeces-shaped object called *Kin no Unko* (Golden Poo), thought to bring good luck, to have sold 2.5 million units. But in the late 1970s, when TOTO turned to relaunching the

Washlet, the toilet – bidet or otherwise – had no place in conversation. It was something detached, unmentionable, out of sight and smell. It could not be advertised. All these factors ensured that the Washlet languished in obscurity for years.

At TOTO, Asuka is joined by Ryosuke Hayashi. His full title is Chief Senior Engineer and Manager of the Restroom Product Development Department, but he prefers to be called Rick, and he is Rick-looking, with slicked hair and almost good English. Rick is an important man. Of the 1500 patents that TOTO has filed in Japan (and 600 internationally), the Restroom Department is responsible for half. Rick finds my interest in the Washlet quaint. It's been around since 1980, after all, when TOTO revamped the Wash Air Seat and launched the Washlet G series (the G stands for 'gorgeous'). I say that for any non-Japanese person used to a cold, ceramic toilet that does nothing but flush, the Washlet is extraordinary. He's unconvinced. I'm asking him about the cathode ray when he wants to discuss micro-robotics.

He'd rather talk about the Neorest, TOTO's top of the line toilet and, in his engineering eyes, an infinitely superior combination of plumbing and computing. Certainly, the Neorest looks gorgeous. It should, when it retails in Japan for $1700, and in the US for $5000. Rick thinks that's value for money, considering that 'it has a brain'. The Neorest takes two days to learn its owner's habits, and adjusts its heating and water use accordingly. It knows when to switch the heat off and which temperature is preferable. It has sensors to assess when the lid needs to be put down, or when the customer has finished and the nozzle can be retracted. It can probably sense that I'm writing about it.

The Neorest's bells and whistles, even if they are nano-technological bells and warp-speed whistles, are vital, because competition in Japan's toilet industry is unrelenting. In 2005, TOTO teamed with the construction company Daiwa House to build the

Intelligent Toilet, which can measure blood sugar in urine, and by means of pressure pads, weight. They have developed the top-secret CeFiONtect, short for Ceramic Fine Ionizing Technology, which uses a super hydrophilic photocatalyst to repel dirt. This complicated procedure is helpfully translated for me as 'like a duck'. Asuka demonstrates the duck glaze properties on a display Neorest in the showroom, marking with a blue pencil both a glazed and unglazed part of the toilet bowl. She looks profoundly unimpressed when the pencil mark is indeed eradicated on the treated area, either because she's done it before or because it's not mascara.

All this technology has come from years of research, billions of yen, many great minds (TOTO has 1500 engineers) and a visit to a strip club.

I persist in asking about the genesis of the Washlet and how it changed Japan, and Rick finally humours me. To sell the Washlet to an unwelcoming public, it had to work properly. The Wash Air Seat and the early Washlet operated mechanically. It took several minutes for the spray to spray and for the water to heat. TOTO solved this by making the workings electronically operated, the spray instant and the angle perfect. The Washlet nozzle extends and retracts at exactly 43 degrees, a position precisely calibrated to prevent any cleansing water from falling back on the nozzle after doing its job (this is known as 'backwash'). Determining the angle was a long, careful process, says Rick. I ask him how the research was done. He says, 'Well, we have 20,000 employees,' and stops. I wait for enlightenment.

Asuka hands me another comic book by way of an answer. It is a forty-eight-page TOTO history published by *Weekly Sankei* magazine in 1985, five years after the company had re-launched the Washlet. Its heroes are Mr Kawakami, a TOTO engineer, and his portly, cheery colleague, Mr Ito. Kawakami and Ito are entrusted with improving the Washlet. The nozzle has to be accurate, and to make it so, they need to know the average location of the human anus. Facts like this are not easy to find, so they turn to the only source material available, which is anybody on the company payroll. Their workmates aren't impressed. 'Though we are colleagues,' one says with politeness, 'I don't want you to know my anus position.'

But Kawakami and Ito prevail by performing the *dogeza*. This is an exceedingly respectful bow that requires someone to be almost prostrate. It is the kind of bow, my translator tells me, 'that a peasant would do to a passing samurai if he wanted the samurai not to kill him'. She says it is an extremely shocking thing to do in the context of toilets. Yet it worked. Three hundred colleagues were persuaded to sit on a toilet – in private – and to mark the position of their anus by fixing a small piece of a paper to a wire strung across the seat. The average is calculated (for males, it comes to between 27 and 28 centimetres from the front of the toilet seat), but that's only the first hurdle. Mr Kawakami is now tasked with improving the Washlet's ability to wash 'the female place'. He needs to know how many centimetres separate a female's two places, and is initially at a loss. Obviously the best place to research female places is in a place with females, preferably naked ones. That's where the strip club comes in, though most strip club clientele are unlikely to react as Mr Kawakami does, by shouting, 'Three centimetres!'

I had fun having the comic strip later translated out loud in a quiet restaurant in England one lunchtime when ears wagged and heads tried not to turn. But the strip club and the wire only go so far in explaining TOTO's extraordinary success. I wanted a second opinion.

Inax is TOTO's arch-rival. The two companies sell similar products, and in fact Inax launched a Washlet-type toilet before TOTO. But it currently has only thirty per cent of the market. The Inax factory is near Nagoya, home of Toyota. I had been given instructions by email to take a slow train from Nagoya to Enokido, where I would be met. The train gets emptier and emptier, and the views more rural and less concrete – pretty curved roofs; barns; gardens – until finally I'm the only person left in the carriage, and we have arrived at Enokido, which is deserted. I don't have directions from the station to the headquarters, so I don't know what to do, until I turn round and see that the station is in Inax's car park. Of course it is. I bet Toyota doesn't have a station in its car park, or its name spelled out in 109 tiny toilets (I counted) on the factory lawn.

I wanted to come to Inax because I'd read about its Shower Toilet. Even in the realm of wonders that is Japanese toilet technology, a toilet in a shower sounded intriguing. A young PR man called Tomohiko Satou has persuaded four senior staff to meet me, and when I tell them this, they laugh. 'Oh, we have that problem,' says Tomohiko. 'The Shower Toilet is called that because it uses a shower – meaning spray – to clean. In the US, we had to call it Advanced Toilet.'

The Shower Toilet is the Inax Washlet, but with a difference. Twenty-seven degrees of difference. Inax has spent a lot of money deciding that a nozzle aimed at a seventy-degree angle has greater firing power and accuracy. They think it cleans better. 'TOTO doesn't want backwash,' says Mr Tanaka, the senior toilet engineer. 'That is why they have forty-three degrees. We don't worry about that because the nozzle is cleaned after every use.' The 1967 version of the Shower Toilet is displayed in the factory showroom. It has a red pedal which had to be pumped to bring up hot water and a blue pedal for cold water. It didn't sell because it cost the price of a new car and with all that water, things got rusty. It was hard to manufacture, with a thirty to fifty per cent ceramic defect rate. Today the defect rate is five per cent.

Mr Tanaka invites me to lunch before a quick factory visit. The cafeteria reception features a perplexing display of a Satis – Inax's luxury toilet and Neorest rival – encased in a perspex bubble in a fishing net, surrounded by shells, sand and blue glass and accompanied by the slogan, 'Our gift to the future'. Tomohiko doesn't know what it means either.

The factory is hot. Inax's ceramic-firing furnace is 100 metres long and burns at 1200 degrees Fahrenheit. The temperature must remain constant, and the factory works almost year-round, because it takes too long and costs too much to fire up the furnace again. The Inax men show me robots that glue and glide beautifully, and which can be trained to do other gliding tasks in only two months at a punishing cost that cannot be divulged. My hosts ask if I have any questions about the production process, but I can't think of any. I'm more interested in the means of consumption than production, and specifically, how TOTO managed to vault over Inax in sales of the high-function toilet – and to convince the Japanese to use it in the first place – when Inax's product was earlier and by some accounts better.

Oh, they say. That's easy. The answer to both questions is the same. It was the gorilla and the actress.

TOTO won over the Japanese public in several ways. On the one hand, there was the gradual approach. Washlets were installed in hotels, department stores, anywhere the public could try them, like them, and never not want to have their bottom washed and dried again. This ensured a slow but steadily growing popularity.

Then came the advertising. In 1982, Japanese television audiences were treated to the sight of an attractive young woman, her hair and clothes slightly wacky – traditional Japanese wooden shoes, a flouncy dress, hair in bunches – standing next to a toilet and telling viewers that 'even though it's a bottom, it wants to be washed too'. The actress was a singer called Jun Togawa, described to me as a Japanese Cyndi Lauper, and she made her mark. Any Japanese who was sentient in

1982 can probably still recite her catchphrases, which were certainly unlike any others: in another ad, she is shown standing on a fake buttock reading a letter supposedly from her bottom, which writes that 'even bottoms have feelings'.

The Inax men sigh. 'TOTO had such good ads. Everyone remembers them.' The Inax ads, by contrast, featured a man dressed up in a comedy costume. 'It was a gorilla sitting on a toilet bowl. It was supposed to be a true experience.' Until now, my hosts have mostly exuded a quiet gravity. Toilets in Japan are a serious business. But the gorilla cracks their composure. They laugh, partly from bewilderment, as they attempt to explain why using a gorilla to sell a toilet could ever have been a good idea. 'We don't know why we had the gorilla,' says Inax's senior communications executive. He has been nodding politely for most of the meeting, but the gorilla story unearths a lovely giggle from inside his composure. 'We can't even remember the slogan. But I do remember that he was wearing dungarees.'

Helped by Japan's economic growth spurt in the 1980s, and by Inax's inept advertising, sales of high-function toilets began a slow, steady climb, but with TOTO in the lead. By 1995, twenty-three per cent of Japanese houses had some kind of Washlet, according to a Cabinet Office survey, and by the end of the next decade, the figure had doubled. Inax has yet to catch up.

The gorilla also failed because the actress hit the right weak spot. TOTO's genius was to address the *wabi sabi* soul of the Japanese consumer. *Wabi sabi* is a cultural and aesthetic philosophy that resists translation, but is usually rendered by the words 'simple' or 'unfinished'. The Japanese tea ceremony is *wabi sabi*, as are those clean bathing habits. The Washlet wasn't unfinished, nor was it transient, but it purified both the body and the toilet room. The toilet was now inside the house – and sometimes inside the bathroom – but its nozzles and hot air kept the user safely distant from his or her bodily excreta. All that complicated engineering simplified the unpleasant business of going to the toilet. Rick Hayashi of TOTO has a toilet-related definition for *wabi sabi*: 'clean, simple, no smell'. The bidet-function toilet removed the need to touch the body with toilet paper. In an increasingly overcrowded urban environment, it provided the means for keeping a distance from bodily functions that before had been achieved by siting the privy far from the house. Also, it had heated seats. It had music. It turned the 4Ks stinky, dark, smelly toilet room into a sliver of pleasant private space, a highly desirable thing to have in the notoriously tiny apartments of Japan's cities.

After five hours of my questions, Mr Tanaka shyly offers two of his own. 'Why don't English people want high-function toilet? Why is Japan so unique?'

I don't know how to reply. I say something vague about how in the UK and US, it's generally presumed that plumbing technology has evolved as far as it needs to. It works, it flushes, and that's all that is required. I say I think that's mistaken, but that's the way it is. Mr Tanaka nods with politeness, but neither of us finds my answer satisfying. I decide to go elsewhere for enlightenment. TOTO and Inax both covet the enormous Chinese market, but what they really want are Americans. US consumers have more wealth and higher levels of technology. In the eyes of the high-function toilet industry, the US is frontier country, yet to be conquered, persuaded and bottom-cleansed. I can't yet answer Mr Tanaka's question, but the American Promised Land might.

TOTO opened its first US office in 1989. Its current premises in New York City are in downtown SoHo, in an expensive-looking building in an expensive location, with an expensive toilet – the latest Neorest – in the window. Somehow, the Neorest is glossy and streamlined enough – it recalls the sleekness of a luxury yacht – to fit in well on this street of designer shops and lofts. The location makes sense because of TOTO USA's business strategy, which is to sell luxury. That's why I'm in SoHo and not Wisconsin (home to Kohler, America's toilet market leader) or New Jersey (home to American Standard, the runner-up).

TOTO USA's PR chief is Lenora Campos. Her manner is assured and her background educated: she holds a PhD in 'the representation of clothing theft in early modern Britain' and describes herself as 'a failed academic'. Somewhere along the way from academia to Neorest, she has developed a talent for euphemism: she describes her job as 'working in high-end plumbing' and excrement as 'matter'. But she's as sharp as her euphemisms are soft. I have come to her with prejudices. The US market is stagnant. American toilets are ugly. They are the 'complex and ridiculous thrones' described by the philosopher Alan Watts, who knew Japan and found Western toilets wanting. Americans aren't interested in innovation, and they don't want Washlets or change.

Campos doesn't bite. TOTO USA isn't only about Washlets. Their regular, non-bidet toilets sell well, though nowhere near Kohler's sales. Campos describes her chosen industry as 'very dynamic. It addresses sustainability, the environment, technology, design.' She disagrees with my interpretation of the industry as dull and conservative. There has been innovation, even if it was only in the plumbing. The innovation is all in the plumbing. Actually, in recent history, this has been the industry's only innovation, and one that was forced upon it.

For decades, the average American toilet used a guzzling 13 litres of water in every flush. Some used nearly 18 litres. By the early 1990s, when several states were reporting water shortages, and the concept of water conservation began to take root, someone noticed that the American toilet was using nearly half a household's water supply. In 1992, the Energy Policy Act (EPAct) was passed, requiring all new toilets within two years to flush with no more than 6 litres. It was a shock. This was barely enough time to change production lines, let alone reconfigure a toilet design that depended on a set volume of water to function. The resulting, modified toilets were rushed and flawed. The six-litre flush had existed in Europe for years, which probably explains its inclusion in the EPAct. If the Europeans can do it, so can Americans. After all, Americans believe that their plumbing is the best in the world (and that Europe's is dreadful); that their sanitary appliances, in the words of anthropologist Francesca Bray, who taught a class about toilets at the University of San Diego, 'are at the top of the evolutionary and civilizational scale'.

But American toilets are nothing like Europe's, and not because they are superior. The American toilet is siphonic, or wash-out. The technology involves complicated principles of air and water flow, but in essence, the US toilet pulls the water out, and the European one pushes it. Manufacturers attempted to make a siphonic flush work with less water by narrowing the pipes, so the siphon effect was increased. It didn't work. Users were having to flush two or three times. There were difficulties with smell. 'In retrospect,' a toilet designer tells me, 'it was pretty asinine to think they would just adapt.'

In plumbing, the post-EPAct era is still known as the time of clogging. Black markets sprang up in old-style toilets. News crews crossed into Canada to interview Americans smuggling back Canadian 13-litre toilets. These toilet pirates were outraged that not only were they being told how much to flush, but they were being asked to do it with bad equipment. It offended their plumbing and their pride. One cross-border black marketeer interviewed by CNN fumed that 'I never thought in Vietnam, you know, when I had to go out in the woods at night, I never thought I'd have a problem here in my own country. [...] We have the best life in the world and we can't even get a decent toilet now.' And anyway, if the new toilets had to be flushed several times, where was the water conservation?

In 2001, enough Americans were angry enough to persuade Representative Joe Knollenberg of Michigan to introduce H.R.1479, the Plumbing Standards Improvement Act. The bill would rescind the low-flow requirements of the EPAct and 'get the federal government out of the bathroom'. It was defeated by one vote in committee.

The clogging reputation was hard to shift. Even today, most American toilets will have a plunger nearby, no matter how much American toilet manufacturers protest that they're out-dated. When American Standard launched their high-end Champion range of toilets in 2003, its selling point was its powerful flush. Posters in faux Soviet revolutionary style featured plumbers in overalls brandishing wrenches, and the slogan 'Working towards a Clog-Free Nation'.

American manufacturers' loss was initially TOTO's gain. TOTO's success in Japan had come through clever advertising and marketing, but it was also due to a brown, gloopy material called *gi ji obutse*, which translates as 'fake body waste'. It is, TOTO staff in Japan tell me, 'a key part of TOTO', and so key, the recipe is top secret, though they will reveal that it involves soybean paste (miso).

Soybean paste is a lethal weapon in the battle for toilet market victory, because toilet makers need to test flushes, and they need test media to do it with. A flush is a chaotic event. Various media bounce around trying to get through one small opening. The more realistic the test media, the closer its properties – buoyancy, density – to human faeces, the better the flush. Toilet engineers have always known this: when George Jennings' Pedestal Vase won a gold medal at a Health Exhibition in 1884, it successfully flushed ten apples, one flat sponge and three 'air vessels' (crumpled paper), as well as cleaning 'plumber's smudge' smeared on the toilet bowl surface.

By the time EPAct came into force, American manufacturers had barely progressed from the apples. They worked with golf balls, sponges, or wiggly bits of plastic. TOTO, though, had been working with a realistic test media for over eighty years. When the National Association of Home Builders (NAHB) published a survey in 2002 testing toilets for flush performance, TOTO models were ranked first, second and third. This helped TOTO's reputation and sales: since 2003, annual US sales have doubled (from 14.4 billion yen [$120 million or £61 million] to 30.1 billion yen [$257 million or £131 million]). TOTO won't release sales figures – beyond saying unhelpfully that the company is 'the recognized leader in the toilet category', which would puzzle Kohler – but at least temporarily, *gi ji obutse* helped to give them the flushing edge in a clogged nation.

Suddenly, America's plumbing industry found it had to catch up. Money was put into innovation. In 2002, American Standard had no PhD's in its R&D department, and now it has five, including an expert in nanotechnology (used to develop anti-microbial coating). But American toilet manufacturers still needed a better test media. They couldn't risk clogging when their reputation was already battered in the eyes of a plunger-weary public, and they could hardly offer their toilets for test drives. Luckily, one day, a Canadian named Bill Gauley got suspicious.

Gauley is a water engineer by training and curious by nature. By the 1990s, six-litre toilet models had gone on sale, after Canadian states brought in water-efficiency rules, but Gauley was sceptical. He did some tests and found that many of the six-litre models were actually using several litres more. When the NAHB report was published in 2002, he read it carefully. The report was supposed to help municipalities choose which toilet models were efficient enough to deserve rebates from the government. Dozens of toilets had been tested using sponges and paper balls as test media, and then rated with scores.

Gauley emailed the NAHB and told them politely that their survey was useless. He said they should have used realistic test media – since when did humans excrete sponges? – and that their scoring system was flawed. 'To their credit,' he tells me, 'they said, "You sound like you know what you're talking about, so raise the funding and you can test the toilets yourself." Then I had to put my money where my mouth was.' His first challenge was to find something superior to sponges. He tried potatoes, mashed bananas, flour and water. Nothing floated or flushed the same way that human faeces did. He read that TOTO used soybean paste and asked them for the recipe. When the company refused to reveal it, he asked his colleagues for help. Anyone who went shopping was instructed to 'look for anything that might work'. They brought back rice paste and peanut butter, but still Gauley wasn't satisfied. Finally someone brought in a brand of miso that he thought looked and floated right. 'Not that we go round feeling human faeces, but some of us have kids and it seemed right, for density and moisture content.'

All that remained was to set up a drop guide to guarantee the test media always fell in the same spot. (Gauley did this electronically, rather than enlisting the help of his colleagues' anuses.) Also, he had to calculate the weight of an average deposit. This wasn't easy, as most research focused on unusual diets, but a 1978 study in the gastro-enterological journal *Gut* eventually yielded the fact that an average bowel movement weighed 250 grams (roughly half a pound). Then Gauley started testing. Of forty toilets that supposedly conformed to the six-litre requirement, only half passed. The results were published as the 'Maximum Performance (MaP) Testing of Popular Toilet Models', and shortly afterwards, the phone calls began. Some manufacturers were furious. Lawyers were consulted. Gauley was not intimidated. 'We'd video-taped every test. So when they came threatening to sue, we'd show them a good performing toilet and they would usually say, "You're right. We have to improve our toilets."' And Gauley had to improve his test media. The soybean paste was the right density and weight, but it was messy, and it wasn't reusable. Then a technician said, 'Why don't you just put sand in a condom?' The physical properties of sand are nothing like faeces, but the comment gave Gauley an idea. He bought a packet of Lifestyles non-lubricated, and returned to the lab. His colleagues were doubtful. 'They said, are you sure it's going to be strong enough?' He filled one with miso and threw it against the wall. It was strong enough.

After TOTO's secretiveness, I didn't expect Gauley to reveal the recipe of his *gi ji obutse*, and in fact, he's contractually forbidden from doing so. When he found the right brand, he asked to buy 250 kilograms from the importer. 'His eyes lit up and he said, "How many restaurants do you own?" I said none and that actually he'd think it was funny but I wanted to use it to test toilets. He didn't think it was funny and suddenly he didn't want to sell it to me any more.' Gauley changed the importer's mind by promising never to reveal the name of the company. But he plans to publish the recipe online once they've analysed it. 'I'm always thinking, how can we help the marketplace? I don't want the recipe to be proprietary. I'm not trying to sell artificial poo.'

Thanks to Gauley's artificial poo, Veritec's MaP is now the best-known independent survey of American toilets available. It is fair to say he's helped make America's toilets better, though Pete DeMarco, a senior toilet man at American Standard, keeps his praise on a low heat. He calls MaP 'one test among many'. In fact, he says, a strange macho one-upmanship has taken over the male arena of toilets and testing. To pass the MaP test, toilets have to flush five of the 250g condoms and four toilet-paper balls compiled of six sheets of toilet paper each, but some manufacturers go further, bigger, stronger. American Standard's toilets are made to flush 1000 grams. This bigger-better mentality has reached the consumer. 'People want 1000 gram toilets,' says Gauley, wonderingly. 'But even 500 grams is a waste of performance.' An interior designer friend says clients still ask her for thirteen-litre 'traditional' toilets, not understanding that a successful flush uses the force and flow of water, not just volume.

Gauley says the marketplace has changed 'incredibly' since he started playing around with soybean paste. I ask him whether the place of the toilet has changed in American culture, whether it has risen above its basic function. He says no-one has ever asked him that before but now that I mention it, no. 'Americans want one that works and then they want to forget about it. And that's it.'

Ironically, the flush transformation brought about by better test media was bad for TOTO. Gauley's tests helped other manufacturers reach TOTO's flushing standards. The company had to find another way to conquer the American market. So it would go back to bottoms. In Japan, TOTO successfully sold its toilets on the concept that they could keep the consumer clean, rather than the other way round. It would do the same in America.

In 2007, the expensive 'Clean is Happy' campaign was introduced to the American public. There were smiley-face badges handed out on the street, viral internet ads and a lavish website featuring disturbingly cheery people telling you what Washlets could do in language Americans could understand. The deodorizer, one cheery person explained, 'is kind of like the catalytic converter in your car'. It is 'a hands-free clean', said another. It uses water, and what's so scary about that, when 'we wash our faces and hair with water! Humans love water!' I was doubtful. American humans may love water, but not to clean their backsides with.

On the website of the American Bidet Company, company founder Arnold Cohen, who prefers to be called 'Mr Bidet', expresses his conviction that the bidet 'is the most significant innovation for personal hygiene and sanitation since the introduction of indoor plumbing'. But the bidet has known limited spread beyond its French origins and even in France, it is disappearing. Ninety per cent of French homes used to have a bidet; now it's ten per cent. Yet if logic governed human cleansing habits, the bidet would be as common as the toilet. Instead, it has generally been viewed with suspicion or bewilderment. (One American schoolteacher visiting Paris in 1929 wrote in her diary, 'oh what a mistake we made about the little bathroom for the feet or what not'.)

As Alexander Kira writes, the bidet entails 'somewhat special circumstances surrounding the cleansing of the perineal region [that are] in some instances, highly charged emotionally'. New York University sociologist Harvey Molotch, who has written about toilets as consumer items, thinks the bidet has never risen above being seen as unavoidably French, and therefore louche. For centuries, Paris was the place to go for sex and women. Anal washing meant dirty naughtiness, something that may have inspired one American manufacturer to name its bidet model Carmen. The abyss between paper and water was highlighted at a 2005 art show held in New York called *Lota Stories*, in which Americans recorded their experiences of using a *lota* cup of water in their toilet habits. The results revealed years of frustration. One contributor, mindful of the frustration of trying to use water in the toilet-paper world of America, left useful advice for subterfuge. Filling a plastic cup (preferably khaki, black or 'some other nondescript colour') at the sink will draw less attention. In an apartment-sharing situation, always keep a plant in the bathroom to explain away the watering can. Above all, use discretion. 'Ignore the impulse to explain what you are doing, even to friends. Unless people have been using a *lota* all their lives, the benefits completely escape them, and they will view you as a freak with a freakish bathroom custom.'

There was another problem. To sell its cleansing products, TOTO had to tell Americans they were dirty. Its first attempt didn't start well. A huge billboard ad featuring bare bottoms, supposed to hang near

Times Square, had to be modified when a church in the building under the billboard successfully applied for an injunction. Bare butts, said Pastor Neil Rhodes, would impede churchgoers' concentration. 'You have naked bodies before your eyes,' he told the *New York Post*. 'How are you going to close your eyes and seek God?' The ad was an odd move to make in a country where conservatism can border on the puritan. Lenora Campos of TOTO is sensitive to this. 'Americans do have issues around the body and bodily functions. We are very uncomfortable discussing it.' The billboard was changed because, she said, it was 'off the mark. If the message is being lost and something is being generated that is unforeseen, then that message has to be changed.'

Delicate sensibilities have always made selling toilets and toilet products difficult. It's hard to advertise your product when social mores don't allow you to say what the product is for. Toilet paper manufacturers have responded to this in mostly uncreative ways (except for the 1920s slogan, 'Ask for Hakle and then you don't have to mention toilet paper'). But since then, toilet paper advertising has been unrelenting pastels and puppies. It's dull, but it works. The global toilet paper industry is worth $15–$20 billion, and according to the most recent statistics available, the average American uses fifty-seven sheets a day.

In 2002, the toilet tissue brand Velvet departed from the norm by launching a campaign that featured 'a series of lovingly photographed bare bottoms', with the tag line 'Love your bum'. It became the second most complained-about ad that year (the first, an image for an anti-poverty charity, featured a cockroach emerging from a baby's mouth). The world of toilet paper, said a creative director for Velvet's ad agency, 'had a huge gap' compared with the creativity levels of advertisers dealing with other markets.

Toilet advertising in the US was in equal difficulties. American

Standard's Soviet-style campaign was successful because it was unusual. But most advertising still featured conservative shots of the classic American 'throne' toilet, stiff in its lines and defiantly un-streamlined. At American Standard, the throne has been modernized by making it even higher, the better to take the strain off ageing baby-boomers' legs. It's now an astonishing 16.5 inches from rim to floor, even more ergonomically nonsensical than usual (squatting frees up the colon and aids defecation; sitting squeezes it shut and impedes release, leading to claims that the sitting toilet has contributed to increased rates of colon cancer, haemorrhoids and constipation). Even with all the flow dynamics and nanotechnology, the modern American toilet has actually only perfected the removal of waste from the toilet while impeding the removal of waste from the body. And the American public is happy with it.

TOTO hopes to sell its products on their health benefits. Colonic irrigation is increasingly fashionable; why not another form of healthy cleansing? But toilet paper manufacturer Kimberly-Clark also tried to appeal to health concerns when it launched Cottonelle Fresh Rollwipes, moist toilet paper on a roll. In surveys, two-thirds of Americans polled agreed that moist tissues cleaned better than dry paper. Kimberley-Clark consequently spent $100 million on the launch. Sales of Rollwipes were dismal, and the concept disappeared from shelves. It has yet to be resurrected. Americans apparently don't want water anywhere near their perineal region, at least not yet.

Consequently, TOTO is playing the celebrity card. When Madonna visited Tokyo in 2005, for the first time in twelve years, she proclaimed publicly that she'd missed the warm toilet seat. Celebrities who have admitted to owning Neorests include Jennifer Lopez, Will Smith and Cameron Diaz. As it did in Japan, TOTO is trying to create toilet evangelists who will do the informal marketing work. When the $1.5 billion Venetian Resort in Las Vegas was being built, TOTO products were placed in all its bathrooms, probably because its billionaire chairman Sheldon Adelson had been given Neorests to test in his home. If TOTO USA can't achieve the mass conversion it did in Japan, it will take the high road of exclusivity instead. You won't find TOTO in a Home Depot, even though that's where you'll find most toilet-buying Americans.

It took fifteen years for TOTO to be successful in Japan. That's the usual amount of time for new household products – air conditioners, washing machines – to be widely adopted. There are signs that Americans may yet succumb to the robo-toilet: in 2007, the American toilet market leader Kohler thought the market was robust enough to launch its own toilet with bidet attachment. Campos thinks the increasing visibility of the toilet in popular culture will help. She cites bathroom scenes in *Sex and the City*, in which the character of the uptight lawyer Miranda is disturbed by her boyfriend's habit of peeing with the bathroom door open, and in *Friends with Money*, Jennifer Aniston is shown cleaning a toilet (though she actually had a toilet-cleaning double). Campos says that the Neorest is starring in an upcoming film, even if its role is to perform the old foreigner-gets-wet story, which hardly seems good advertising. Celebrities such as Will Smith and Barry Sonnenfeld, director of *Men in Black*, have spoken out against the deficiencies of toilet paper. (Sonnenfeld compared using a moist wipe to 'a romp through a field of daisies for your butt'.)

Perhaps the robo-toilet revolution is simply taking its time. But Tomohiko Satou of Inax is noticeably lacking in TOTO-style optimism. He has a fair sense of American views about robo-toilets, having spent time posted in Inax's San Francisco office, where sales, he admits, were 'not so much'. 'Japanese people,' he tells me, 'understand that our product is very sanitary and clean.' But years of trying to explain that to Americans taught him a painful truth. 'Americans just don't want to use it. They're not scared. They're just not interested.'

Where would you hide? **German Toilet Organization exhibit, featuring Jack Sim, among others** – *German Toilet Organization*

3. 2.6 BILLION

'... and sanitation'

From my post in the first-floor café, I have a clear view of the proceedings below. A small man with black hair is speaking with authority to a dozen or so reporters. Behind him is the usual architecture of a convention, except the booths hold toilets, a fact which doesn't stop Russian beauties in scanty clothes draping themselves over urinals and steel sinks. Between the booths, babushkas wander, their faces unfazed under their patterned headscarves.

Jack Sim is launching the latest summit of the World Toilet Organization in the Manezh conference centre in Moscow, a skip away from the Kremlin, and he is doing it with his usual weapons of humour and grave facts. Two-thirds of the world without sanitation. Over a billion people drinking filthy water every day. If the world can talk about eating food (he calls this 'uploading'), then isn't it time they talked about downloading too?

The reporters take notes and record as required, and the next day the results are what I have come to expect. Bad, tired puns abound.

Flushed with pride. Pooh-poohing the competition. Raising a stink. One newspaper op-ed, though supposedly in praise of public conveniences, couldn't resist sniping that the WTO was 'unaware perhaps of the rather bigger group of the same name based in Geneva'.

Of course it wasn't. Sim knows exactly what he's doing. He chose the abbreviation to get the kinds of coverage he now does get, even if it's only usually around World Toilet Day on November 19, or at the WTO's two annual conventions. Sim is a humorous man, and his jokes – some smutty, usually funny – are fired constantly (he tells one reporter that both WTOs deal with big and small businesses), but his vocation is sombre. To get the world to talk properly – or at all – about sanitation, any weapon will do. 'People joke about it,' he says, 'but when the jokes stop, they listen. When they've stopped listening, they take action.' And action is certainly required. The second half of the twentieth century saw billions of dollars spent on trying to provide safe and healthy methods of excreta disposal to developing countries. And still the numbers were horrifying and still the projects failed. By the end of the 1990s, sanitation needed help, and it got Jack Sim.

Sim makes me think of a sprite. It might be his height, which is not enormous, or the shiny black hair falling in a floppy fringe, or the crackling, zinging energy. I notice the zing because he tells me his story at eight o'clock one morning on a coach travelling through Moscow's outskirts. The World Toilet Organization delegation is on a field trip to see the Russian Toilet Museum – actually a field of Portaloos – and I'm about to doze when Sim bounces into the seat next to me to continue a conversation we'd started two days earlier. I'm not at my best in the morning hours, so I switch on the recorder and switch off my brain. Sim doesn't need any stimuli. He has his own.

In 1996, he was a successful businessman running a mid-sized company that traded in building materials. He had an agreeable house, a wife, four children and the trappings of success. His home city of Singapore had been transformed from a middling island into a flourishing business centre of skyscrapers and gloss, where cleanliness and order were prized enough for long-haired rockers to be banned along with chewing gum. Then one morning, Sim read in the newspaper an appeal by the Prime Minister for an increase in 'social graciousness'. 'He was talking about the cleanliness of public toilets, and he was blaming users' behaviour. It could have been true, but what causes bad behaviour? If I go to a toilet and it's smelly, I don't think I'll behave very well.'

He was offended by the article, but it made him think. He began to look at toilets. They were indeed in a shameful state. By now his business could run smoothly without him, so he could take on a new project. Why not toilets? I ask him whether he started reading up on it. 'No,' he says, firmly. 'What for? I had forty years of experience to consult from.' He remembered watching other kids in his poor neighbourhood running around with worms hanging out of their backsides. He thought of all the filthy toilets he had visited in his home country, despite its wealth and development. He concluded that there were problems and that taboos weren't helping. He thought 'there was a need to legitimize the subject'. In 1998, he formed the Singaporean Restroom Association and began plans for a global organization. Using $150,000 of his own money, he travelled around the world to learn what he needed to, but from experts, not from books.

First he checked out what was already happening in the sanitation world. There wasn't much, despite the number of high-profile public commitments that have been issued over the last three decades. The 1980s had been designated the International Drinking Water and Safe

Sanitation Decade by the UN. By the end of it, poor toiletless people were supposed to have safe, adequate sanitary facilities. But they didn't, and the decade was extended to the end of the 1990s and renamed the Third Water Decade. Since 2005, the world has been in the grip of the decade of Water For All. There have been high-profile water-related conferences in abundance, though sanitation is always an afterthought, if it's considered at all. (At the 2004 World Water Week, the biggest global water event, sanitation was relegated to a seminar held before the conference even began.)

After all these conferences and dedicated decades, the world was left with four in ten people still defecating in fields and on roadsides, and with diarrhoea still killing more children under five than HIV, tuberculosis or malaria. There were some hopeful signs: one in three of the world's people who gained toilets got them between 1990 and 2004. Pete Kolsky, a senior sanitation specialist at the World Bank, thinks this is worthy of praise. 'When you think of the long history of humanity – e.g. 10,000 years of urban living – that strikes me as an unsung achievement.' Also, with population growth 'it's very hard to stay in place'. All targets are moving ones.

By the 1990s, investment had declined along with momentum. The authors of the 2006 Human Development Report, an annual publication of the United Nations Development Programme, wrote that 'when it comes to water and sanitation, the world suffers from a surplus of conference activity and a deficit of action'. Several factors contribute to this ineptitude. Responsibility for sanitation was fragmented: twenty-three UN agencies have some responsibility for it, but no agency takes the lead. There is a United Nations Office for Outer Space Affairs, but none for resolving the biggest public health crisis on this planet, and one which the Human Development Report authors put in its proper place. 'The 1.8 million child deaths each year related to clean water and sanitation dwarf the casualties associated with violent conflict. No act of terrorism generates economic devastation on the scale of the crisis in water and sanitation. Yet the issue barely registers on the international agenda.'

Outside the UN and development world, there was a similar lack of focus. There were well-established toilet associations – in Britain and Japan, for example – but they were doing their own thing. The Japan Toilet Association installed 10 November as National Toilet Day (as 11/10 in Japanese sounds like the characters for 'clean toilet') and ran good conferences, but in Japanese. The British Toilet Association set up a successful Loo of the Year competition (successful enough that winners' revenue allegedly doubles) but its members came from the plumbing industry and weren't desperate to solve the world's sanitation problems. There was no global, organized association campaigning to improve the world's rotten sanitation state. Jack Sim thought the planet needed one. In 1999, he founded the World Toilet Organization. He knew what he wanted it to be: a support network for all existing organizations. It wouldn't charge membership fees. It would be, in his words, 'a servant, not a leader'.

Sim also wanted the new WTO to be a campaigner. He wanted to create a global lobbyist for the cause of better and sustainable sanitation, because there was none. But he had never been an activist. He wasn't sure how to go about it. He needed a mentor. He was Singaporean Chinese, from a tradition that respected wisdom and experience. Instinctively, he did what many sanitation professionals now propose: he decided to learn from an HIV campaigner. The similarities were clear: both HIV and excreta-related diseases arise from activities considered private and unspeakable. So Sim went to meet Mr Condom, the name by which Mechai Viravaidya, a former Thai senator, is best known.

Viravaidya is an impressive man who speaks fluent English, having grown up in Australia. He wears a bow tie and tweeds and has an idiosyncratic wit that must have endeared him to Sim. During parallel campaigns to reduce Thailand's population, Viravaidya persuaded women to use the Pill by calling it 'the family welfare vitamin', and having village shopkeepers sell it. Thais considered condoms a taboo topic of conversation, along with sex, but Mr Condom was undaunted. He launched a 'cops and rubbers' campaign, got taxi drivers to distribute contraceptives, and opened the Cabbages & Condoms restaurant in Bangkok, on whose tables a bowl of condoms take the place of the after-dinner mints. (Its name derives from Viravaidya's theory that condoms should be as cheap as vegetables.) He also enlisted the army in his crusade, persuading military radio to broadcast safe-sex messages.

One lesson that Sim learned was about language. Viravaidya tailored his message to his audience. He held condom-blowing contests with schoolchildren, and talked profit and loss with businessmen (his punchline was 'dead customers can't buy anything'). 'He's not talking to them about high morals or anything like that,' Sim tells me with admiration. 'He's talking to them in a way they like to hear.' Partly thanks to Mr Condom's efforts, new HIV infections in Thailand decreased by a staggering eighty-seven per cent.

Sim came away from meeting Viravaidya having learned that anything could be made talkable. Also, 'you have to laugh at yourself first, because people are going to laugh. But after they laugh at you, they will listen. If I hadn't been taught that, I would never have gone any further than a couple of months.'

Instead, his World Toilet Organization has now run eight World Toilet events attended by over 4000 people. I go to two events, and they are both memorable, though Bangkok beats Moscow for flair with a line of toilets on-stage, their lids spelling out Happy Toilet, and by being the only WTO event to be held under martial law. (The Thai military had kicked out the Prime Minister a few months earlier but judged a toilet conference no threat to national security.)

The Moscow summit nonetheless has a certain glamour. The

Manezh conference hall is a gracious building only slightly marred by a large portable toilet parked outside as part of the expo. Downstairs, a fur exhibit is packed. Upstairs, the toilet show attracts a smaller crowd, despite the pull of a terrorist-proof toilet whose concrete foundation makes it safe from suicide bombers, apparently. It's not a bad selling point: toilets are vulnerable places, especially for royals. Henri III of France was supposedly stabbed by a monk while at his business in 1589, and the fifteenth-century Scottish king James I was murdered by noblemen while hiding in his privy.

WTO events can produce a colourful cast list. There may be a Russian professor of hydraulics, who appears at a breakfast meeting with a can of beer in his pocket. There is usually a delegation of Chinese who give short presentations and are not seen again because, according to other delegates, 'they only come for the shopping'. There are always things to learn. During breaks, I find out Australian automatic toilet operators use the terms Code Red (for cleaning incidents involving blood) and Code Brown (for the obvious). The Australian automatic toilet operator who tells me this is Scott Chapman, whom I meet in the first-floor café. He looks bored. He is a broker by profession, but has been reluctantly roped into the automatic toilet business by his father, a man conveniently called W.C. Chapman, for whom he is standing in. He wants to know why anyone would possibly want to write a book about sanitation, and looks astonished to learn how much of the world doesn't have any, and of the consequences. We kept in touch over the next year, and I watched with amusement as he turned from reluctant toilet delegate to a sanitation-solving pioneer. He went to tsunami-hit areas and learned about septic tanks and groundwater contamination. He became seduced by vacuum toilets, because they use little water and can be installed with regular sewage systems. He learned, as I learned, that not taking the WTO seriously is easy and a mistake.

Behind the jokes, Sim is resolutely committed to his cause. (This is a man who has named his four children Faith, Truth, Earth and Worth, because 'I gave them names with virtues so they could build on them'.)

Thus far, the WTO has set up a World Toilet College in Singapore. There are plans for a Peace Prize for Sanitation, a rock concert in China, and for a Toilet Development Bank that would give $100 low-interest loans to encourage poor people to build latrines. Over the months, I get emails informing me of yet another of Jack Sim's systems, and though he is a showman, his systems generally come true, because he's a networker. On the WTO website, the organization's logo – a blue toilet seat – is now featured alongside the familiar blue laurel wreath of the UN because the WTO is beginning to be given proper weight by the development establishment. Plans are being made to set up a Global Sanitation Bond, a finance mechanism which is meant to improve on existing ways of getting money from rich countries to poor ones.

All these activities, Sim says, means the WTO is proceeding according to plan. 'The right way to do any social work is to work until you're not needed any more. I think this is the way to do it: to do evangelism; to get legitimacy in the media and to have a united global voice. Then the political process will come. I try to market the subject like a religion.' Religion needs a charismatic high priest, so Sim plans to turn to Bollywood or Hollywood. 'I think for us to come to terms with the fact that we go to the toilet would be quite easy. It only needs a few movie stars to talk openly about it.'

Talk to anyone who is trying to improve the world's sanitation, and this idea will become a refrain. We need a champion. A Bono or a Geldof. A Nelson Mandela or an Angelina Jolie. A film star or a politician who has the courage to talk about toilets, when most people only want to talk about taps.

The Netherlands-based International Water and Sanitation Centre recently listed celebrities who do charity work for water. Hollywood star Matt Damon has launched the NGO H2O, whose mission is to 'bring clean water to Africa'. In the music world, the rapper Jay-Z did a three-part series for MTV on the world's water crisis. The rock-singer Chris Martin is an ambassador for WaterAid. Not one celebrity ambassador, however, has made the obvious point. Dirty water is usually dirtied by faeces. It is hard to supply clean water when that clean water is contaminated by overflowing pit latrines or filthy fingernails. But water is blue and fresh. Sanitation is not. 'We can get celebrities to talk about water,' a WaterAid employee tells me. 'But none of them want to be pictured on a toilet.' Clean water gushing from a new hand-pump makes for great press coverage. Accompanying a child to her new latrine does not. WaterAid probably isn't fussy. If there's no Bono or Bob Geldof, they'll take a regular politician who champions sanitation. Except there aren't many of those either.

There are some subjects that politicians just don't like, and sanitation is among them. A sanitation expert with several decades of experience tells me he can list the influential political figures in sanitation on one hand. He rethinks. Half of one hand. 'Maybe politicians will run on a platform of water for all or houses for all,' he says. 'But you've never seen one run on a campaign of no shit for all or sanitation for all.' They think it won't bring votes. Voters don't generally complain about poor sanitation. Even the ones with sewage in their living rooms stop protesting once the stink has gone. This is mistaken in all sorts of ways. In the Indian city of Pune, for example, a local councillor who had been trying for years to get latrines for his slum ward was overjoyed when they were finally built. 'The next two elections are in the bag for me,' he said. 'All I have to do is stand outside the toilet with my hands folded and a smile on my face.'

People with decent sanitation have fewer diseases and take fewer days off work; they don't have to pay for funerals of their children dead from cholera or dysentery. They save on medicines, and the state saves because it's not providing expensive hospital care. Every dollar invested in sanitation brings an average $7 return in health costs averted and productivity gained. That simple number is the result of years of complex calculation of variables by development economists. But Joseph Chamberlain, Mayor of Birmingham, knew in 1875 that lost wages and medical costs due to preventable disease came to £54,000 a year, two to three times what it would cost to build sanitary accommodation. Globally, if universal sanitation were achieved by 2015, it would cost $95 billion, but it would save $660 billion.

Such economic theory has been bolstered by real-world proof. When Peru had a cholera outbreak in 1991, it cost $1 billion to contain but could have been prevented with $100 million of better sanitation measures. During the first ten weeks of the epidemic, losses from agricultural revenue and tourism were three times greater than the total money spent on sanitation during the previous decade.

Sanitation is one of the best investments a country can make. It can benefit health, education, productivity and tourism revenues. But its all-round benefits contribute to the problem. Excreta-disposal is a political football, kicked between departments. A cartoon calendar published by the World Bank's Water and Sanitation Program (WSP) shows a young girl dressed in brown who looks anxious. She is Sanitation. Three bureaucrats stand nearby, one each from the Ministry of Water, Health and Local Government. 'She's your responsibility,' says Water. 'No, you take her,' says Health, and Local Government echoes him. And Sanitation holds out her hand hopelessly, surrounded by flies.

One obstacle to progress is the way the advantages of improved sanitation are calculated. Health economists measure benefits by something called a DALY (Disability Added Life Year). This is supposed to calculate the value of a life 'saved' – actually prolonged – by a health intervention. Averting a child death is worth 30 DALYs, for example. But most sanitation benefits are more life-changing than life-saving, and earn fewer DALY points as a result. A $100 latrine can't compete with a 20-cent bag of Oral Rehydration Salts (used to combat diarrhoea) on a strict cost-benefit calculation.

Eddy Perez of WSP says these priorities are reflected in government aid budgets. He tells me that there is 'almost zero money' for water and sanitation in the budget of USAID, Washington's largest aid donor. 'They still fund some water and sanitation but it comes out of their environment budget which is much smaller. And a lot of it has to do with politics and reconstructing areas that they've blown up, like Iraq. The health sector of USAID does not see investment in sanitation as a public health investment. That's a huge problem.'

Even in countries that have a water supply and sanitation ministry, 'watsan' – the development short-hand for water and sanitation – always trumps 'sanwat'. Water always comes first. Governments already dedicate only 0.5 per cent of their budgets to watsan, on average. (Pakistan, for example, spends 47 times more on its military budget than on water and sanitation, though it loses 120,000 people to diarrhoea a year). Of that, most goes on water. This ordering of governmental priorities causes what Darren Saywell of the International Water Association calls a capacity constraint problem. 'The best and the brightest of people generally don't end up in your environmental sanitation department. It's not seen as a particularly clever career move.' People talk of the twin sectors of 'wat' and 'san', says Perez, but that's a misnomer. 'They're not twinned by any stretch of the imagination. One's tall and healthy and the other is short.' When Perez gives presentations, he shows a slide of Arnold Schwarzenegger and Danny DeVito from the film *Twins*. He doesn't have to spell out who represents what.

It takes a rare politician to see the economic sense of good sanitation. The one that everyone talks about, four years after he left office, is Ronnie Kasrils, former Minister for Water Affairs in South

Africa. He's the most successful sanitation minister ever, a UN official tells me. He's the only minister who was 'willing to stand up and say that sanitation is where it's at,' says Darren Saywell. A South African friend describes him as 'quite a character'. All South Africans know who Ronnie is, he says, because he was a guerrilla commander for the African National Congress (ANC), and consequently exiled for decades. Kasrils is now South Africa's Minister for National Intelligence, or its top spook. I had to meet him.

In South Africa, sanitation has been political for decades. During apartheid, water and latrine supplies were distributed according to colour. White settlements, even poor ones, usually got flush toilets and waterborne sewerage. Blacks had bucket latrines – a bucket with a seat on top – or nothing. By the time democracy came to South Africa in 1994, a relatively rich country had, for reasons of racism and politics, desperate sanitation figures comparable with those of poorer African countries with fewer resources. The new government gave the responsibility of sanitation to the Ministry for Water Affairs and Forestry, but mostly got on with more pressing matters like leavening racism, alleviating poverty and trying to run a new, wounded country.

President Thabo Mbeki appointed Kasrils to Water Affairs in 1999. He was already a well-known figure. A Jew from Johannesburg, he joined the South African Communist Party after the Sharpeville massacre in 1960, then Umkhonto we Sizwe (MK), ANC's armed wing. He was instrumental in blowing up some electricity pylons around Durban and went into exile in 1963. Over the next two decades, he spent much time in guerrilla camps in the bush. These years, he likes to say, are what prepared him for his sanitation future. When the ANC was elected in 1994, he became Deputy Minister for Defence. He didn't expect to be moved to water. He was a soldier and knew about soldiering. But soldiers know about sanitation because they have to. Shit can win and lose wars. Accounts of the battle of

Agincourt describe half the English archers fighting while naked below the waist, because dysentery was so ravaging their troops (this led Voltaire to conclude that England had 'taken victory with its pants down'). During the First World War, France's general staff ordained that latrines be painted light blue, because this was the colour that flies liked least. In Vietnam, the Viet Cong laid thousands of sharpened wooden stakes topped with faeces – *pungi* – and caused thousands of casualties (the stick pierces the boot and foot; the excreta is deadly).

So Kasrils didn't have trouble seeing the connection. He joked that when he was a guerrilla, he had known the rivers and forests of the whole region. He said, 'I'm going from fire to water.'

We meet in the Ministry for National Intelligence, which is an unsignposted building off a highway leading into Pretoria. There is little visible security. The staff, who have the quiet courtesy common to powerful corridors, don't bother checking me for recording equipment or cameras or anything else, and Kasrils seems equally at ease. He is a handsome sixty-year-old, courtly for an ex-guerrilla, and wearing a dark pin-striped suit and waistcoat. He refers to himself as 'one', and I wonder if one did that in the bush.

I'm surprised that he agreed to meet, when he's the equivalent of the head of MI5 or the CIA, who I'd never get to see. There's also the fact that he no longer has anything to do with sanitation. But South Africans tell me he loves press attention – he published an autobiography called *Armed and Dangerous* – and also, he loved his past job. And why shouldn't he, when he got to be photographed sitting on toilets?

In the beginning, the new Minister for Water Affairs did what water ministers do: he concentrated on the wat and didn't pay much attention to the san. South Africa's water problems were certainly dire: millions had no water supply thanks to decades of discrimination, and 'service delivery' became the political demand of the new South Africa. So Kasrils got water to people, and only changed tactics when cholera broke out in KwaZulu Natal in 2000. He had an epiphany about what he calls the human waste factor. 'You see that people's water supplies are contaminated, and you see that it's because of pollution, so you start to analyse it and you get to human waste.'

This awareness led him to the Water Supply and Sanitation Collaborative Council (WSSCC), a Geneva-based UN agency run under the auspices of the World Health Organization. At the time, it was directed by an Indian engineer called Gourisankar Ghosh and chaired by a forthright Briton called Sir Richard Jolly. Ghosh came to South Africa. The minister and the UN's top man for excreta disposal found they had much in common. Both men, like Mechai Viravaidya in Thailand, were trying to speak about the unspeakable with people who didn't care to listen. 'No-one wants to talk about shit, do they?' says Kasrils. 'I'd go to *mbizos* [public meetings] and no-one ever says, "You know, man, I'm sick and tired of this disgusting latrine I've got."'

The same applied in Cabinet meetings. But after the cholera outbreak, Kasrils began to bang his new drum. Soon, he'd banged it loud enough to be nicknamed Minister for Toilets. In league with Ghosh and the WSSCC, he began to strengthen networks, such as an African alliance of water ministers. The ministers knew each other already, but 'whenever we came together we'd talk about dams and water'. Sanitation was an afterthought, if anything. He changed that. 'Whenever we'd go through documents and come to something about water blah blah, we'd always add "and sanitation".' With a strong network in place, they could go global. Kasrils agreed to host a sanitation forum in South Africa called AfricaSan. It would be held a few weeks before the Earth Summit – the UN summit on Sustainable Development – that would convene in Johannesburg in August 2002, and its job would be to get sanitation noticed.

At this point, and to this day, the development and aid world was obsessed with the Millennium Development Goals (MDGs). This set of eight goals had been agreed upon at the UN Development Summit in 2000. The MDGs are not modest. They commit signatories – 191 countries at last count – to do things like halve the number of people living on less than a dollar a day (Goal One); educate boys and girls equally (Goal Two); halt and begin to reverse the spread of HIV/AIDS, malaria and other major diseases (Goal Six); and all by 2015. But when the MDGs were announced, there was a notable absence. HIV killed fewer children than sanitation-related disease, but sanitation was nowhere to be seen. An impact needed to be made. At the AfricaSan conference, a video was played. It showed an old man washing the hands of a young girl. It was nothing that hadn't been seen on a thousand UNICEF videos. Then the camera pulled away and the old man was shown to be Nelson Mandela, who said, 'Now we must all wash our hands.' In the words of one audience member, the effect was 'Wow. Bang.' (That a not very creative video could be 'Wow' showed how stagnant and unloved the sanitation sector felt.)

The effect was strengthened by colourful photo opportunities, including one that featured Richard Jolly and Ronnie Kasrils seated on toilets brandishing toilet paper. Stunts, admittedly, but they made sanitation visible because no-one had done such things before. Behind the scenes, Kasrils and others 'argued like the blazes' to add sanitation to the MDGs. The Americans didn't want any more targets, but eventually agreement was made, and Target Seven was added to Goal Ten. This enjoined the world to 'reduce by half the proportion of people without sustainable access to safe drinking water'. 'Access to improved sanitation' was added as Indicator 31. Yet the targets that have to be reached for Indicator 31 – those 2.6 billion toiletless – are larger than the targets for HIV, safe water, malaria or nutrition.

There is no doubt the MDGs are flawed. Darren Saywell of the IWA calls them 'a useful political tool, but not a professional one'.

Reducing the number of people who don't have something means knowing how many people don't have something in the first place, but only 57 out of 163 developing countries have counted the poor more than once since 1990. Ninety-two have never counted them.

Yet the anointing of sanitation as being worthy of UN and MDG attention was important. In the world of development and donors, money was freed and some priorities changed. Two years later, the WSSCC capitalized on the revamped MDGs by launching probably the best media campaign about sanitation and hand-washing that has been seen since the Lever brothers persuaded a grubby Western world to think of soap as a vital necessity. The WASH – Water, Sanitation, Hygiene – campaign was sharp and clever. In development terms, it was called advocacy. I call it good advertising. There were posters and postcards with smart slogans, such as 'Hurry Up! 2.6 billion people want to use the toilet'; or 'One billion people have a drinking problem'. There was poetry such as:

Jack and Jill went up the hill
To fetch a pail of water
After a drink of the water
Jack died of cholera
and Jill died from amoebic dysentery.

Kasrils began to champion WASH. He liked its messages. 'Cholera and typhoid,' he tells me, 'kill so many million kids a year, which amounts to two jumbo jets full of children crashing every four hours.' In the years after 9/11, this was a powerfully vivid image to use. Kasrils thought the WASH campaign provided him with great soapbox material. 'They were giving me gems. I jumped to it.' He describes those heady sanitation times as 'tremendous fun. The best period of my life.' It was also busy. When Kasrils took up his post, 18 million South Africans lacked sanitation. The government's target was to eradicate this 'backlog' by 2010. By 2003, Kasrils' ministry was delivering 85,000 toilets a year. It would need to deliver 300,000 a year to meet the 2010 target, but it was a start. Also, it meant a lot of toilets to be officially opened.

Kasrils leaps out of his chair at this point. He wants to tell me a story, and because he's a dramatic man – before he became a guerrilla, he did some theatre and worked in advertising – he needs to act it out. A village had installed Ventilated Improved Pit latrines (VIPs), and Kasrils was invited to open the first one, ceremoniously. Persuading a village to adopt VIPs was already an achievement, because after the end of apartheid, black South Africans reasonably wanted what the whites had. They wanted waterborne sewerage, which was high-status. It was also expensive and totally illogical in South Africa, a largely water-stressed country that can't afford, financially or environmentally, to let everyone flush dozens of litres of water down a toilet.

The VIP latrine was a version of the Blair latrine, invented in Zimbabwe in 1973 by Peter Morgan, a British-born engineer who was working with the Ministry of Health's Blair Research Laboratory. There are endless ways to build a pit latrine well and endless ways to build them badly. Millions of the people who count in statistics as having access to adequate sanitation actually have a dark and stinking fly-infested box. The VIP innovated with an offset pit that could hold an interior vent pipe, a screen on the pipe to keep out flies and a semi-dark interior to achieve the same effect. It was definitely ventilated, and definitely improved: a three-month experiment in 1975 found that 179 flies a day were caught in a latrine without a vent pipe, while the daily fly toll in a VIP was two. (If flies can't get into the latrine, they also can't emerge from it with faeces-covered feet, ready to infect nearby food.)

In the village of Umzinyathi, near Durban, Kasrils was invited to inaugurate a villager's new VIP. By now, his thumbs in his waistcoat pockets, he is acting the part of the mayor from a film whose title he can't remember, but which involved the inauguration of a urinal (I think he's referring to the French novel *Clochemerle*, made into a BBC series in 1972 and which my mother remembers as being 'gripping'). 'So the mayor gets out of his car and with real pomp goes into the urinal, and it's one of those where you can see the person's head and you hear the urine and people are waiting and then he comes out.'

In the real-life South African version of this tale, Kasrils duly went into the toilet under the gaze of a crowd of serious onlookers. He shut the door, pretended to use it, and came back out, whereupon the entire crowd began singing the South African national anthem. 'And the VIP had been painted in national colours, of course,' he adds. It's a great story, and I've enjoyed the show, because Kasrils is another great persuader. It's a shame Jack Sim came to toilets after Kasrils had left them, because they would get on.

Kasrils had a lot of persuading to do, on all levels. 'You have these mayors and councillors who don't want to know about [the VIP], because in their mind, if you don't have a toilet and wash it away, it's dirty. They think in terms of the long drops they've been in, where there's flies and buzzing and a stink. You have to break down those barriers.' But waterborne sewerage was out of reach of most municipal budgets. One of the new government's most popular measures was to put into writing that water is a human right, and to provide each citizen with twenty free litres a day. But after that, everyone must pay. 'So you tell them if they have VIPs they won't be spending money, and that it takes ten litres of water to flush a toilet, and that those whites with the toilets? You'll go into their houses and you'll find they're not flushing because they don't want a high water bill.' It's all about psychology, says Kasrils. 'You take steps and you make efforts to mobilize and educate and then it clicks.'

During his tenure, the Minister of Toilets accomplished great things. He persuaded President Mbeki to open toilets, a considerable achievement when South Africa's head of state is rumoured to be prudish. South Africa may be the only country in the world where sanitation was properly addressed, and HIV – given President Mbeki's notorious inability to countenance it – was not.

The president moved Kasrils to the National Intelligence Ministry in 2004 and the sanitation world is poorer for it. His successors have held the title of Minister for Water Affairs, but they have never really been Ministers for Toilets. People refer to Kasrils as 'a glimmer of hope' in the sanitation world, but a glimmer doesn't last. I admired Kasrils' efforts, but he was the political exception in a world that hasn't figured out what the rule is. At least for now, the battle for better sanitation isn't happening in cabinet meetings. As Jack Sim says, the evangelism comes first. The political process will follow. The evangelism is being spread outside powerful corridors and under the political radar, by the ground troops of sanitation. They are foot soldiers who don't mind wearing filthy boots to tramp through possibly the most unappealing public health crisis in the world. Into these ranks, Trevor Mulaudzi – a South African geologist who prefers the nom de guerre of Dr Shit – is fully conscripted.

One day Trevor was driving as usual through the gold-mining areas north of Johannesburg. He worked then as a geologist for Anglo-American, an enormous mining company whose salary provided him with a large house, two cars and a pleasant lifestyle.

As he drove, he saw a group of children on the street. High-school age. 'I stopped them and said, "Please go back to school." The children were amazed, and said they could not go back to school because they were looking for a toilet.' Now it was Trevor's turn to be astonished. '"There's no toilet in your school?"' and he went to have a look, marching into the high school gates and to the ablution block, handily – for an interloper's purposes – set apart from the school buildings. The toilets were a disgrace. 'Shit everywhere! Shit piled up behind the door! Filthy! There were no doors on the stalls. There was even poo in the handbasins.' He finally understood, he says, 'why our children hate going to school. It starts in the toilet.'

The headmaster then received a visit from a strange man who told him his school toilets were disgusting. 'He was astonished. Then he said, "But the children are unruly. They do not clean."' OK, said Trevor, 'I will do it for you.' Then, wearing a suit and tie, he found a wheelbarrow and a shovel and set about cleaning the block with children and teachers looking on at 'this madman who is cleaning our toilets'. He said 'it was an amazing feeling'. So amazing that he went home, quit his job and nice lifestyle and the next day set up a cleaning company whose mission was to ensure that South Africa's schoolchildren had clean toilets.

That's the tale. I hear it several times over the course of our acquaintance; first during a presentation at a WTO event, where Trevor's infectious laugh and trilby kept the audience rapt. Then in a radio interview that Trevor plays as we drive through the streets of Johannesburg on the way to his home. The presenter is suave, but he sounds like most media people who deign to look at sanitation sound. Indulgent. Trevor says he is a 'toilet activist', and that he is lobbying for clean toilets in the country. 'And what do people say?' asks the presenter.

'They say I'm mad.'

'Well,' says the smooth voice. 'It certainly sounds like it.'

Trevor is not mad, but he employs exaggeration deliberately. His wife Audrey, an assured and brisk medical doctor, amends the story somewhat when I ask her about it. 'It wasn't twenty-four hours,' she tells me one afternoon at their house in Linden, a previously white suburb, where the Mulaudzis live with two of their three beautiful and smart daughters (the third is studying in Cape Town and preparing to be South Africa's first female president). 'Trevor was very unhappy at his job. It's true that he came home and said he wanted to quit but it wasn't so cut and dried.' They considered business opportunities and decided on cleaning. 'It was so difficult for five years. The cars were nearly impounded. But the twenty-four-hour thing? That's Trevor.'

When I question him on the epiphany story, he laughs, a lot. 'That's marketing! It's the same reason I go to WTO events.' Promotion helps business, and his business helps to improve South Africa's toilets. He told the Moscow audience that his profits were $30,000 a month. It got a round of applause, but it was more massaging of the truth. In fact, his wife says, he takes money from his regular cleaning business – which cleans supermarkets and mining hostels – to do the school work. It's rarely enough, and they are not rich.

The day after I arrive, Trevor takes me to Khutsong. He used to live in nearby Fochville, a town that seems very white and very blond, even now. I wonder how it was to live here and be neither of those things. Oh, it was fine, says Trevor, except for the time he tried to retrieve a tree branch from his neighbour's garden and she called the police to report a trespasser, though she knew Trevor well. 'What can you do? We have to live with these people.'

Khutsong is a dusty, dull, poor township. (When I ask someone to explain the difference between town and township, he thinks, then says, 'A township is where black people live.') Khutsong has a reputation these days because of riots that began in 2006. The rioters have punched out the traffic lights and burned down councillors' houses and the town library. The riots were about poor municipal services. As usual, sanitation was not mentioned, though when I see what Khutsong's facilities consist of, I wonder its residents didn't riot years ago.

I've asked Trevor to show me a bucket toilet. I can't believe it's actually just a bucket. He knows his old maid in Khutsong had one, but when we stop to look, she's upgraded to a brick-walled latrine. 'Good for her,' says Trevor, and tries the house next door because he's sure she's an exception. An old man arrives and says he's the grandfather of the three children who live here alone, because their parents have died of AIDS. 'Children are bringing up children all around here,' the old man says, his tone flat. At the school nearby, half the 1000 pupils are AIDS orphans. I begin to see that people have other things to think about than toilets, and why the three children have a bucket toilet that is smelly and horrible, and rarely emptied as it should be. Sometimes the municipality trucks come, sometimes they don't. No-one tells me what happens when they don't, but the field opposite says it all.

'That's nothing,' says Trevor. 'Let me show you what schoolchildren have to live with.' We detour up to a nearby primary school to meet a friend of his. Victor is deputy headmaster of Wedela, a primary school for the children of mine employees. He is a gentle man who grew up in Soweto in a house with two rooms and eleven children. Perhaps the harsh circumstances made him more receptive to Trevor, now a close friend, but once just a man who turned up one day and bothered him about the toilets. He'd never given them any thought before. 'Everyone thought it was normal. No-one knew any different. Then Trevor came and I realized how important it was.'

In his memory, the state of things wasn't so bad, but Trevor says he has pictures. There was filth and excrement everywhere. It wasn't the worst Trevor has seen: he carries round a photograph of a school latrine made from a metal car chassis, its edges lethal. But it was bad enough to require several days of cleaning and the use of a heavy-duty pump.

Now, the toilets are pristine, partly because one pupil's mother has become a volunteer toilet attendant. Susan tells us it's a tolerable job and better than nothing. Sometimes she gets tips. We compliment her on her cleanliness and give her 200 rand, an average monthly salary. (Afterwards, Victor tells me that Susan had gone straight to the headmistress after we left and told her that I had drunk out of the sinks because the place was so clean. He thinks this is hilarious.)

Trevor wants me to see what he usually deals with. We go to Khutsong's high school. It looks in good enough condition from afar, but inside it is shabby and missing things. We step over a drain and Trevor makes a sound of disgust or maybe despair. 'Look,' he says, kicking the air where a grate should be. 'They've stolen it! They steal everything.' He is angry. Of course Khutsong is poor, but poverty doesn't excuse everything.

He tells me about his grandparents, who walked to the diamond-mining town of Kimberley, miles from their village in Venda, to get some education, then came back and founded the village school. His father is also fastidious and proud. 'He puts on a suit and tie every day to go nowhere.' From his father, Trevor gets a love of old-fashioned values and hats. He won't find either in this school block, where we wander around hunting for filth. It's easy to find. In the girls' section on the second floor, there are stalls with no doors, toilets that clearly haven't flushed in months and maggots. There are plastic bags on the floor which Trevor says are used for wiping, and condoms. A cardboard box on the floor serves as a 'she-bin,' as South Africans call it, for sanitary towels and tampons, and on it someone hopeful has written, 'Fold your pad nicely and put it in this box, please beautiful children.'

Most of the toilets are broken, so children have used the floor. When that gets too dirty, Trevor says they will go the nearest available facility, which is usually in a *shebeen*, a township bar, and they'll stop for a beer while they're there. Or they'll wait and suffer. The consequence is that children hate school. This enrages Trevor. 'They know that once they get in the school yard they've got no place to pee. You can imagine if you are driving and you want to pee, you can't concentrate any more. Now how about listening to a teacher when you feel like that?' This is not more Mulaudzi marketing. School attendance and sanitation have been linked in dozens of studies. UNICEF estimates that one in three girls in sub-Saharan Africa drop out of school, either when they're menstruating, or permanently, because of poor sanitation facilities. In Tanzania, India and Bangladesh, when schools installed decent latrines, school enrolment increased by up to fifteen per cent.

Trevor is fond of saying that he doesn't promote cleaning toilets but clean toilets. Cleaning is a one-off; clean toilets require sustainable effort. He tries to persuade schools to change their habits by doing toilet espionage; he sneaks in and takes photographs, then uses them as moral leverage to shame the staff into action. But it has only worked in his home province of Venda. Ten Venda schools now sell toilet paper to raise cash to keep their toilets clean. They sell it for two rand a roll and get fifty cents profit. The shock and awe of Trevor's naming and shaming technique can be transformed into calm sustainability, with some imagination and effort.

Trevor is a showman, as much so as his former Minister for Toilets, but he has a serious cause. His business card carries the slogan, 'What does your toilet say about you?' He tells me that he is on a divine mission. 'If I'm the only one who knows why our black children aren't going to school, what will I do on the day of judgment?'

But his mission is difficult. Over the week we spend together, he seems increasingly frustrated. His method of working now has a name: social entrepreneurship, a concept popularized by the American economist Bill Drayton, who runs a foundation that funds people who want to combine business with charity. Social entrepreneurs have been getting more attention in recent years (Mechai 'Mr Condom' Viravaidya, another social entrepreneur, was recently awarded a prestigious $1 million prize by the Gates Foundation), but their influence has yet to penetrate government offices. Trevor needs access to education planners, but he claims he can't get any. He says that because he's not a charity, there's no place for him in government contracts. So I take Trevor with me on a round of official sanitation appointments. He says, 'I'm playing the white woman foreigner card!' and that he would never get through the door otherwise.

In fact, he gets on winningly with Portia Makhanya, Director of Sanitation in Kasrils' former Ministry of Water Affairs and Forestry. This is not difficult: Makhanya is a warm and chatty woman who used to be a mental health nurse. She is a Xhosa from the Eastern Cape, and when I ask her if she can sing the only Xhosa song I've ever heard of – Miriam Makeba's 'Click Song', which uses the clicks of the Xhosa language – she does so gracefully and beautifully. She also does a fine impression for us of a person trying to take a shit in the bush with a pig nearby waiting for her to finish. For a while, she and Trevor reminisce. First about the township uprisings in 1970, when a policeman was given a choice of being killed or eating from the nearest latrine. (He chose the toilet option.) Then about their childhood latrines made from corrugated tin. So hot, hot, hot! They talk of the Amabhaca, men from the Bhaca minority tribe who emptied your family bucket latrine and were treated as untouchables. You remember? You stayed out of their way, because they were angry men, with angry scarification on their faces. And you never shook their hands.

A ten-minute appointment with Portia turns into a two-hour chat, and it's fun. She is honest about the realities of trying to get sanitation to 15 million people. South Africa's targets are amusingly ambitious. Bucket latrines are to be eradicated by December 2007, though there are 132,000 still in 'established settlements', and at least the same number again in illegally settled areas. Nearly five hundred clinics are supposed to get sanitation by the same date, and all schools will be provided with decent latrines. But not a single clinic toilet has been built, the money has yet to be released, and our conversation is taking place six months before deadline.

To be fair, Portia's mandate was made more difficult when a decentralization programme handed sanitation implementation to local governments. Her ministry was left with deciding norms and setting goals. She says with some weariness that 'we try to influence local governments but we no longer have teeth'. This is another truth of sanitation; to reform it in most countries, entire systems of governance have to be changed. In South Africa, the situation is worsened by a serious lack of skills and capacity, after years of apartheid, and reform seems a distant hope. I think of Ronnie Kasrils and I ask Portia whether there's any South African sanitation champion equal to the former Minister for Toilets. Whether there's any glimmer. She smiles sweetly. 'It's still us! With our teeth. With our false teeth.'

On my penultimate day in South Africa, Trevor and I fly to Cape Town brain-squeezingly early on Kulula, yet another cheery low-cost airline. I don't know who decided that low-cost airlines had to try to be funny but they did and they do. The flight attendant says, 'We have landed in Cape Town. If that's not where you want to be, that's your problem.'

We do want to be here, partly to meet Trevor's daughter, the presidential hopeful. We're also here to meet Shoni, an old acquaintance who got in touch after hearing Trevor on the radio. He is a manager at Robben Island, Nelson Mandela's former prison, and gives us free tickets to visit. 'I can't accompany you, I'm afraid,' he says over dinner. 'I have to take the President of Singapore on a tour.'

The next morning, we arrive at Robben Island as the president is leaving. Trevor tells people that he runs the South African Toilet Organization, though there isn't yet any such thing. But Shoni has told this to the president, who knows Jack Sim and the WTO and wants to meet Trevor. The meeting takes place inside a circle of Singaporean journalists, who exude an urgent keenness to record this momentous event, but no seeming distaste or scorn. It is a spectacle of Trevor Mulaudzi marketing, and it is as powerful as ever. We keep the tour bus waiting for ten minutes, and I am embarrassed. I ask Trevor to charm them. He gets on and says, 'Excuse me ladies and gentlemen but that was the President of Singapore and he wanted to talk to me about shit.'

I don't know what to think of Trevor. A week of the Mulaudzi roadshow is alternately impressive and head-scratching. I don't doubt his enthusiasm for his cause. His plain-speaking is refreshing. But six months after I leave, there's still no official sign of the South African Toilet Organization he had promised me he was about to set up. Portia Makhanya had furnished a list of municipal education contacts, but Trevor has made no new contracts with schools.

Instead, I receive a surprising email from him. He writes that Malaysia's Deputy Minister for Housing, head of the country's toilet association, has invited him to the country to sort out its school toilets. So he's going to move to Malaysia. I'm disappointed. What about those photos he shows people of schoolchildren grinning and happy because they have a decent toilet, which he captions 'South Africa's future'? What about Jack Sim's call for more home-grown toilet evangelism? But he is impervious. He's tired. 'In South Africa I am not given the opportunity to do what I do best [which is] cleaning toilets. Our government does not even know me. The toilets in Malaysia are very dirty and the government needs help from professionals like me, Dr Shit. They can't wait for me to get there. They are even giving me Malaysian citizenship, AMEN.'

Mumbai, India – *Author*

4. GOING TO THE SULABH

Spade, Black, Dung, Horse

It drips on her head most days, says Champaben, but in the monsoon season it's worse. In rain, worms multiply. Every day, nonetheless, she gets up and walks to her owner's house, and there she picks up their excrement with her bare hands or a piece of tin, scrapes it into a basket, puts the basket on her head or shoulders and carries it to the nearest waste dump. She has no mask, no gloves and no protection. She is paid a pittance, if she is paid at all. She regularly gets dysentery, giardia, brain fever. She does this because a 3000-year-old social hierarchy says she has to.

In the beginning, the Original Being created four *varnas*. From his mouth came the Brahmins, who would be the priests, teachers and intellectuals. From the arms came the Kshatriya, the warriors and rulers. From his thighs came the Vaisya, who were the administrators, the bureaucrats, the merchants; and from his feet the Being formed the Shudrya, the farmers and peasants. Inside these varnas are thousands of sub-groupings, each with a traditional occupation attached. All of it makes up the Hindu caste system, still pervasive and influential in modern India. In its report *Broken People*, Human Rights Watch summed up caste as 'the world's longest surviving social hierarchy, [...] a complex ordering of social groups on the basis of ritual purity'. It is indeed complex, changing from region to region and from one religious interpretation to another. But all over India one thing is common: beneath the castes are the outcastes, the polluted and the untouchable. They are untouchable because they handle human shit.

They used to be known as *Bhangi*, a word formed from the Sanskrit for 'broken' and the Hindi for 'trash'. Today, official India calls them the Scheduled Castes, but activists prefer Dalits, a word that means 'broken' or 'oppressed' but with none of the negativity of Bhangi. Most modern Indians don't stick to their caste jobs any more. There is more inter-caste marriage, more fluidity, more freedom than ever before, but the outcastes are usually still outcastes, because they are still the ones who tan India's animals, burn its dead and remove its excrement. Champaben is considered untouchable by other untouchables – even the tanners of animals and the burners of corpses – because she is a *safai karamchari*. This literally means 'sweeper' but is generally translated into English as 'manual scavenger', a term popularized by India's British rulers, who did nothing to eradicate the practice and much to keep it going. This scavenging has none of the usefulness of its usual meaning. There is no salvaging of waste, no making good of the discarded. Champaben recycles nothing and gains nothing. She takes filth away and for this she is considered dirt.

There are between 400,000 and 1.2 million manual scavengers in India, depending on who is compiling the figures. They are employed – owned, more accurately – by private families and by municipalities, by army cantonments and railway authorities. Their job is to clean up faeces wherever they present themselves; on railway tracks, in clogged sewers. Mostly, they empty India's dry latrines. A latrine is usually defined as a receptacle in the ground that holds human excreta, but dry latrines often don't bother with receptacles. They usually consist of two bricks placed squatting distance apart on flat ground. There is no pit. There may be a channel or gutter nearby, but that would be luxury. The public ones usually have no doors, no stalls and no water. There are still up to ten million dry latrines in India, and they probably only survive because Champaben and others are still prepared to clean them.

I meet Champaben in a village in rural Gujarat. Like every other state in India, Gujarat is bound by the 1993 Employment of Manual Scavengers and Construction of Dry Latrines (Prohibition) Act, which makes manual scavenging illegal on pain of a year's imprisonment or a 2000 rupee ($45) fine. On paper, Champaben doesn't exist, and on paper, she is as free as the next villager. Untouchability has been illegal in India since 1949, when it was abolished by means of Article 17 of the Constitution of India.

Champaben knows that. But what can she do? Scavengers have been doing their work since they were children, and they will do it until they die, and then their children will take over. Champaben's mother-in-law Gangaben is seventy-five years old. She has been scavenging for fifty years. In a village nearby, I meet Hansa and her daughter Meena, who is ten. Meena has already been introduced to her mother's job, because she has to do it when her mother is ill or pregnant or both. Most manual scavenging is done by women, because they marry into it, and they have no choice. Men in the manual scavenger class often hide their profession from prospective brides until it's too late, and they can then escape their foul work in alcohol, because they have a wife to do it for them. Some scavengers work in cities as sewer cleaners, and unclog blockages with their bare hands, their only protection a rope. They are regularly killed. Last year, three men died of asphyxiation, one after the other, when they entered a manhole in New Delhi.

The women talk freely. They are chatty and assertive and pristine. I look at them and try to see the dirt on them and in them, but I can't. They are elegant and beautiful even when they bend down to pick up the two pieces of cracked tin they use to scoop up the faeces; when they demonstrate how they sweep the filth into the basket; when they lift the basket high with arms glittering with bangles, with considerable grace. Their compound is dusty but not dirty, though they are not given soap by their employers – whom they refer to more accurately as their 'owners' – and though they are not allowed to get water from the well without permission from an upper caste villager. They offer me a tin beaker of water, and the water is yellow. 'Look at it,' says Mukesh, an activist from a local Dalit organization called Navsarjan, who has accompanied me. 'Look at what they have to drink.' The beaker presents a quandary. I consider pathogens and faecal-oral contamination pathways, but also that they'll expect me to refuse to take a drink from an untouchable, because many Indians would. I take a sip and hope for the best, feeling pious and foolish, imagining bugs and worms slipping down into my guts, wreaking havoc.

Mukesh has been to this village before. Plenty of well-meaning activists have been here before. 'You come here all the time, you interview people,' says Gangaben. 'And what do you do? Nothing.' Gangaben is the most indignant. She disappears into the house and returns with two chapattis – flat-breads – on a plate. 'Look at this,' she says. 'This is what I was paid today. Scraps.' Privately employed scavengers usually get paid five rupees (about ten cents) per month, per house. Municipal day wages are thirty rupees (less than a dollar) a day, but scavengers are often unpaid for months on end. Who will dare to stand up to their employer? When I ask Hansa to show me where she works, she refuses. No way. 'My owners would skin me alive.' She is deadly serious, and deadliness is something she has to consider.

*

There are laws to protect Dalits, to criminalize untouchability and to outlaw manual scavenging, but they are not enforced. Violence and abuse against Dalits is endemic and unceasing. Over six months in 2006, the Centre for Human Rights and Global Justice surveyed the Indian media for reports of abuse against Dalits and gathered a few headlines:

> 'Dalit leader abused for daring to sit on a chair'
> 'Dalit lynched while gathering grain'
> 'Dalit beaten for entering temple'
> 'Dalit girl resists rape, loses arm as a result'
> 'Dalit tries to fetch water beaten to death'.

Navsarjan calculates that three Dalits are killed each day. Police statistics from the last five years, gathered by the National Campaign on Dalit Human Rights, tell a similar tale of brutality. According to those – definitely incomplete – statistics, thirteen Dalits were murdered per week and three Dalit women were raped every day. (A human rights worker tells me with bitterness that 'untouchability doesn't apply when it comes to vaginas'.) During my most recent trip to India, I read a newspaper article of a Dalit schoolgirl who had been gang-raped somewhere in provincial India, but I didn't note the name of the village. When I looked for it later online, searching for 'gang rape' and 'Dalit', I got through three pages of results before despondency made me stop, because they all related to other cases.

A 2006 survey of 565 villages in 11 states found that Dalit children in 37.8 per cent of government schools were forced to sit apart from other children during meal-times. Part of that 37.8 per cent is Hansa's daughter Meena, a pretty child who tells me she's not allowed to sit with her school friends. When I ask Meena what she wants to do when she grows up, she puts her head in her hands.

The same survey found that health workers refused to enter Dalit homes in thirty-three per cent of villages, and in nearly a quarter of villages, postmen refused to deliver mail to Dalit houses. Control and prohibitions percolate into all aspects of Dalit life. The Indian writer Gita Ramaswamy quotes some elderly manual scavengers in Hyderabad, who were reluctant to give their names. Then they explained why. 'We were told very categorically by the upper-castes that our names were to be self-ridiculing. If any parent or grand-parent chose a fair name of the child, we were instantly abused for having lost sight of our *aukath* (social and moral position).' They list their names. Jhamta, Kaloo, Gobar, Ghoodo. Spade, Black, Dung, Horse.

Champaben, Hansa and Gangaben have no need of statistics. They need only try to wash themselves at the village water source when they are dirty with shit and they will be turned away by the higher-caste women who are there. They need only try to enter their village temple and they will be refused access. They need only ask for a glass of water from their employers/owners and watch as water is poured directly into their cupped hands, so that no crockery is dirtied. In times not too long gone, Dalits were made to wear bells round their necks to warn of their passage, because even their shadows are polluting. It makes sense to Gangaben. 'We carry excreta on our heads. Of course we are unclean.'

Theories about the origins of the scavenger caste vary. Perhaps it came about when the Mughal emperors used prisoners of war to clean their wives' harems, or perhaps that's handy anti-Muslim prejudice. One of the fifteen duties for slaves listed in the holy text Narada Samhita was the disposal of human excreta. India is not unique in treating people who work with dirt as polluted. The old gong-fermors of Victorian London knew to keep to themselves. One medieval edict forbade baiting of cesspit-emptiers on pain of a fine, and what would be the point of having a fine for something that never happened? In a milder version of marginalization, toilet attendants and sewage workers worldwide admit to lying about their place of work, turning themselves into 'hygiene managers' or 'local government workers' to avoid scrutiny or disgust. The director of a London sewage treatment works admitted to me that when he tells new acquaintances what his job is, 'some people do move three feet backwards'.

Nor is Hinduism the only religious system to set out purity rules. Deuteronomy 23 instructs Jews to 'have a place outside the camp and go out there, and you shall have a spade among your tools, and it shall be when you sit down outside, you shall dig with it and shall turn to cover up your excrement'. The Bible texts found in the Dead Sea Scrolls are more precise: proper hygiene requires Jews to defecate between 1000 and 2000 cubits (1500 to 4500 feet) away from camp, in a northwesterly direction. (Something didn't translate to the New Testament, though, as the Tartars were reported to have a curse that exhorted enemies 'to tarry so long in one place that thou mightest smell like thine own dung like the Christians do'.) One sacred Hindu text calls for Indians to fire an arrow, and defecate only where it lands or further. The Vishnu Purana, dating from the first to the third century BCE, instructs followers to defecate at least 150 feet from a source of water, and to urinate 15 feet away from habitation. The Buddhist text Vinaya Pitaka, a rule book for monks, is expansive in its toilet provisions. Proper Buddhists should, among other things, not defecate in the toilet in order of seniority but of arrival; cough loudly upon arriving at the toilet (and if there is an occupant, he should cough in response); not defecate while chewing tooth-wood; nor grunt upon defecation; and not wipe oneself with a rough stick.

None of this is much comfort to the scavenger women. But they don't expect comfort. They don't expect anything. 'Our caste is written on our forehead,' says Champaben. 'Ours is low and yours is high. That's the way it is.' But a young girl called Dhurmisthu is less entranced by tradition. 'The caste system has nothing to do with religion. It's a conspiracy maintained by the upper castes. We think we're equal, but they just see brooms in our hand.'

Young urban Indians contend that caste is irrelevant now, because India's huge metropolises act as a mixing bowl, diluting old traditions and backward thinking. What happens to Hansa and Gangaben, they will say, is a thing of the villages, of peasants, of tradition and history. They will point to successful Dalit lawyers, politicians, academics. The current Chief Justice is a Dalit. Plenty of politicians in India's upper house are Dalit. Under India's Scheduled Castes reservations system – which is controversial but widely implemented – Dalits benefit from positive discrimination for employment and university places. But they are still Dalits, and there is still caste. Surveys show that the majority of young Indians still expect to have an arranged marriage, and forty per cent won't marry outside their own caste or state.

The glass ceiling pressing down upon the scavengers' heads consists of cultural prejudices, but also of economics. When I first wrote about manual scavengers, for the American magazine *Jane*, the first draft of my story came back punctuated with the editor's questions. She couldn't understand why scavengers felt obliged to do this work, and who employed them. She wrote, 'Who are their bosses? Uneducated farmers? (I'm assuming that the more educated the people are, the less tolerant they are of these conditions.)' That's a nice hope. In fact, manual scavengers continue to be employed by municipal authorities, who use them to clean sewers, and by Indian Railways. Last year, the company declined to say when it could phase out the use of manual scavengers to clean its tracks. Until fully sealed flush latrines were installed on its trains in place of the current 'open discharge' ones, scavengers were the cheapest cleaning option. A high court in Nizamabad only demolished its dry latrine – cleaned by scavengers – when ordered to do so by the Supreme Court.

For Navsarjan, the solution to manual scavenging and untouchability is loud activism and alternative employment. There are programmes that provide manual scavengers with loans to set up small businesses, but only a fraction of their funds have so far been disbursed. A newer initiative will provide more loan money and skills training. It also specified 2009 as the new target for the total eradication of manual scavenging. The previous target had been 2007, and there had been others. Safai Karamchari Andolan, another noisy Dalit organization, prefers direct action over empty promises, and often demolishes dry latrines by hand (the rubble, carefully labelled, is kept in a cabinet in the SKA office). Dr Bindeshwar Pathak, however, preferred not to destroy toilets, but to build them.

Pathak founded the organization Sulabh International in 1970. It is now India's largest charity, with fifty thousand staff. Millions of Indians have installed the Sulabh Shauchalaya latrine. Of more interest to non-Indians will be the half a million public toilets that Sulabh has built all over India. Every day, ten million Indians – and plenty of relieved foreign travellers – use a Sulabh toilet, because they are in railway stations, airports, on the main streets of India's cities. Pathak's toilet blocks are so common, Indians now say, 'I'm going to the Sulabh,' and the word 'toilet' can be left silent.

Sulabh's headquarters consist of a pleasant campus near Indira Gandhi International Airport in New Delhi. The campus has green lawns that refresh after the yellow brown dust and chaos outside, and signs urging visitors to smile, please, because you're in Sulabh now.

I have visited Sulabh twice, and each time the procedure was the same. First, assembly, held in a low hall near the kitchen that cooks with biogas, fermented from the excrement deposited in the Sulabh public toilet complex next door. To reach the meeting hall, you walk over those green lawns, which get their colour from being irrigated with effluent from the same toilets, cleaned with sand filters and UV light. On the way, you will pass through part of the outdoor display of cheap latrine models, the reason Sulabh came into existence, and you might find the odd visitor or two, as I did in forty-degree heat one day in May, when an artist's assistant from London was busy gluing blocks of dried Rajasthani excrement for a show by the Spanish artist Santiago Sierra. (The blocks were shown in a plain room, simply standing, because their substance was deemed subject enough.) After one visit to Sulabh, such encounters seem normal, because this is a place that has a science lab that contains glass jars filled with 'Balls of Dried Excrement', and where deeply courteous scientists in white coats will express their great excitement about the sewage cleansing properties of duckweed, or dip their hands into a box of dry, brown granulated stuff and say, 'Excrement! Like gravel!'

Then you will be led into the hall, presented with a beautiful silk scarf and a garland of flowers, and you will watch while children in neat blue uniforms, the girls with red ribbons in their plaited hair, sing the Sulabh song whose lyrics exhort you to 'come together and build a happy Sulabh world'. These are the children of the Sulabh school, housed in a complex on the other side of a yard of demonstration pit-latrine models. It is a unique school, because two-thirds of its intake are the children of manual scavengers. The education they are given here ensures that their parents' job will not also be theirs too one day. It is a happy scene, but it has been built on forty years of one man's stubborn conviction that scavenging is a sickness in his culture, and that toilets can heal it.

In the late 1960s, the young Pathak committed a grievous sin. He was studying sociology, and like many young Indians getting used to being part of a newly independent and ambitious nation, he was an idealist. His ideals were those of Mohandas K. Gandhi. The father of the modern Indian nation was one of the few political leaders in history to publicly talk about toilets. There is a scene in Richard

Attenborough's biopic where Gandhi argues with his wife because she refuses to clean their latrine. She says it is the work of untouchables; he tells her there is no such thing.

Gandhi also argued with everyone else. At the 1901 Congress Party convention, he told delegates it was a disgrace that manual scavengers were being used to clean the latrines. He asked delegates to clean their own latrines and when they didn't, he publicly cleaned his own. The eradication of manual scavenging was a recurrent theme throughout Gandhi's life. He called the practice 'the shame of the nation'. He wrote that 'evacuation is as necessary as eating; and the best thing would be for every one to dispose of his own waste'.

For a great politician to talk freely about such things in public was impressive, but Gandhi's position had its critics. Plenty of Dalits object to Gandhi's comparing scavengers with mothers looking after children, when the children are upper-caste Indians content to keep these 'mothers' in a state of servitude. The great Dalit politician Dr B.R. Ambedkar, who thought that 'inequality was the soul of Hinduism', wanted caste to be abolished, not tinkered with.

In 1969, the idealistic young Pathak began to volunteer with the Gandhian Centenary Committee in his home state of Bihar. The committee's job was to organize three years of programmes and celebrations in honour of their hero's birth. Their hero cared about scavengers, so the volunteers were supposed to do the same. For a young man from an Orthodox Brahmin family, this was unthinkable.

In his spacious office at Sulabh headquarters, Pathak, now sixty-four, tells a family anecdote. 'When I was a child, I wanted to know why some people were untouchable. I wanted to see what would happen if I touched one, so I did.' His grandmother made him eat cow dung and sand, drink cow urine, then take a ritual bath. How can dung be clean? Purity rituals that seem to defy sense are common to many cultures: Ancient Mesopotamians carried dung around their necks to ward off evil; Hindus have decided that cow dung is holy.

Such classifications of what is dirty and what is pure are obviously not about reality. Investing dirt with power makes it more manageable. Making people defilingly dirty makes them more manageable, too. They can be kept in their place. Even so, writes Virginia Smith in *Clean*, her history of hygiene, 'Distancing yourself from poisons, dust and dirt is one thing but distancing yourself from invisibly "unclean" people and objects is quite an achievement of the imagination.' It was a leap of imagination that Pathak refused to make.

Instead, a few years later, he risked more cow punishment by going to live with scavengers. There, he found both outrage and a vocation. He couldn't believe people lived in such conditions. The state of Bihar had for years been running a latrine-building programme state-wide in an attempt to remove the dry latrines that scavengers had to clean. Yet the women carrying headloads of excrement were still there. 'Scavengers' appalling hardship, humiliation and exploitation,' Pathak wrote, 'have no parallel in human history. [...It is] the utmost violation of human rights.'

Gandhi's tactics of encouraging brotherly love across caste boundaries, and urging Indians to clean their own latrines, had failed miserably. The status quo was too convenient. Pathak decided a better solution was to provide an alternative technology. Scavengers' jobs would never be surplus to India's needs, not with a population of a billion excreting people. Perhaps the solution was to make scavengers unemployable by eradicating dry latrines. Not by knocking them down, but by providing a better latrine model that didn't require humans to clean it, but was cheap and easy. Most importantly, it had to be easy to keep nice. Given a choice between a smelly, dirty latrine and the street, even the most desperate might choose the latter. Pathak read WHO manuals about pit latrines, and developed his own version.

It had to be on-site, because India has neither water nor sewers enough to install expensive waterborne treatment systems. Even today, only 232 of India's 5233 towns have even partial sewer coverage. Indian urban wastewater treatment consisted of dumping it in rivers. The mighty Yamuna River, which supposedly dropped to earth from heaven but actually runs nearly two hundred miles from the Himalayas through the nation's capital, has millions of gallons of sewage poured into it every day. By the time it reaches Delhi, the Yamuna is dead. As for the Ganges, its faecal coliform count makes its supposedly purifying waters a triumph of wishful thinking, unless the purification is the kind you get from chronic diarrhoea, dysentery, or cholera.

Pathak called his new latrine the Sulabh Shauchalaya (Easy Latrine). It was twin-pit and pour-flush. It could be flushed with only a cupful of water, compared with the ten litres for flush toilets. There was no need to connect it to sewers or septic tanks, because the excreta could compost in one pit, and when that was full, after two to four years, the latrine owner could switch to the other, leaving the full pit to compost. This was another Gandhian concept: the Mahatma had used the phrase *tatti par mitti* (soil over shit), and would dig a pit for his own excreta then cover it with soil when it was full. The Great Soul of India was a pioneering composter. The Easy Latrine leached its liquids into the ground but supposedly without polluting groundwater. Most importantly, it was cheap, with the most inexpensive model costing only 500 rupees ($10).

Despite all this, Pathak's technology found no takers for three years. He had to sell some of his wife's jewellery, and resorted to peddling his grandfather's bottles of home-cure remedies. Until one day, when he entered an office in a town in Bihar and sold the idea of the Sulabh model to the municipal officer on duty.

The Sulabh model consisted of more than the latrine. It was also a method. Pathak saw how the aid and grant-making world worked. Budgets and donor cycles are fixed. They can be withdrawn after a few years with little notice. Pathak decided that Sulabh would not accept grants. It would make sanitation a business that paid for itself.

It doesn't sound radical but it was. In the 1970s, development experts were convinced that poor people wouldn't pay for sanitation. Since then, this has been proven to be nonsense. Poor people pay up to ten times more for water – from water gangsters or private tankers – than a resident with municipal water supply. UK regulations concluded that spending more than three per cent of the household budget on water was an indicator of hardship. But poor people in Uganda, for example, spend twenty-two per cent of their budget on water.

Pathak thought people would pay, so he developed a range of models for all budgets and tastes. His social service organization would be non-profit, but it would be a business. This thinking was new.

Gourisankar Ghosh was working as an engineer in India in the 1970s, before he went on to head the UN's Water Supply and Sanitation Collaborative Council. Ghosh thinks Pathak and Sulabh have been revolutionary. 'In the 1970s in India no-one was talking about sanitation. Worldwide not even the World Bank was addressing it. People were suspicious of Pathak because he's a self-made man. So he did it on his own, without UNDP [United Nations Development Programme] or state support. He's a visionary and a pathmaker.'

This may explain the messianic aura that surrounds him. He is treated with insistent subservience by his staff, and described in a history of Sulabh as possessing 'awesome innocence' and 'manners which are compelling'. Powerful men in India are shown deference as a matter of course, but a British sanitation expert who knows Pathak expresses astonishment at 'the amazing cult of personality around him'. In a foreword to the Sulabh history, Pathak writes, 'I do not claim the eminence of Faraday, who invented dynamo, Lazslo Biro, inventor of ball-pen, [...] Einstein, or of that unknown Assyrian who invented the wheels, or of the Caveman who "invented" fire but one thing is common between me and those great names: none of them was an engineer.'

Neither was Pathak. To promote his Easy Latrine, he had to battle for thirty years with suspicious World Bank–influenced engineers who intended to carry on installing what they'd always installed: unsuitable waterborne treatment plants which were prohibitively expensive to run for Indian municipalities, and whose maintenance required levels of expertise that rarely existed. Nonetheless, Pathak's efforts have won awards: in 1992, he became the first man to take the topic of latrines to the Vatican, where Pope John Paul II awarded him a St Francis of Assisi medal. In 1995, he was the Limca Records (the Indian Guinness) Man of the Year. Despite this, writes the Indian journalist S.P. Singh, Pathak remains indifferent to fame and fortune, seeking only to 'rescue scavengers from the tyranny of the social system in which one man's excreta is another man's headload'.

This may be true. But causes cost money, and the Easy Latrine didn't bring in enough revenue to cover Sulabh's running expenses. Pathak decided to build public toilets, too.

In the 1970s, public facilities in India were a rare sight. The few in existence were squalid and offered little advantage to defecating on the pavement outside, so people often chose the street instead. Pathak had an idea that was simple, new, and apparently doomed. If people had a clean toilet with water and light, they'd probably be willing to pay for it. 'People laughed at me,' he recalls. 'They said, in Bihar, people don't pay for bus tickets and rail tickets. Why would they pay for toilets?'

But his negotiation skills served him well, because in 1973 the first Sulabh public toilet opened in Patna, the state capital of Bihar. It had water, electricity and round-the-clock attendants. Sulabh charged one rupee for toilet use, and urinals for men were free (women could also urinate for free, but they have to specify their needs to the caretaker). A wash cost two rupees. In the first day, Pathak says, five hundred people used it.

Gourisankar Ghosh remembers working in 1980 in an office in Kolkata that overlooked a lake. 'One morning I saw a lot of activity and someone told me someone was building a toilet. I little realized that people had no public toilet. I never thought it would work but once it was built I saw women and men coming to use it. I was amazed. It cost 1/40th of a dollar and they were using it happily.'

Sulabh's concept of pay-per-use was not new – a similar government programme had been tried, and failed, several years earlier. But the business model was. Instead of funding toilets with government grants, Sulabh approached authorities and municipalities and suggested something different: if the authority paid for the cost of constructing the toilet and provided the land, Sulabh would run it for a set number of years and keep the profits. The business model was an attractive one to municipal authorities who, back then, could not be bothered with sanitation. 'Before, no-one wanted to know,' says Pathak. 'In the beginning, we couldn't find anyone willing to tender to construct toilets. The upper castes wouldn't consider it. They wouldn't even come to meetings. Now they fight for the tenders. We have blended social reform and economic gain.'

In Mumbai, I take a tour of Sulabh's public conveniences. My host is Chandra Mohan, the head of Sulabh's Mumbai branch. Like other senior Sulabh employees, he began volunteering for Sulabh after retiring from a top business position. Other Sulabh men were high up in the civil service. Good connections help business.

We whizz through Mumbai in a white Ambassador, India's greatest car and a vehicle that manages to feel luxurious even when it's decrepit, for it sits its passengers high and it has lines as noble as its name. All the stops on the bathroom tour are on prime Mumbai real estate. There is a Sulabh next to Mumbai city hall and another by the Gateway of India. Sulabh also has the only establishment operating on the city's famous Chowpatty beach. There used to be others, but Mumbai is cleaning itself up, and they were demolished for aesthetic and health reasons (they were ugly and stank). Some of Sulabh's 750 Mumbai toilet blocks are in slum areas. 'We cross-subsidize them,' says Mohan. 'The ones in high-volume areas bring in money to pay for the ones in poor areas that don't.'

He takes me to a public toilet near the headquarters of *Indian Express*, a prestigious weekly magazine. It is well-kept and pristine, unlike many government-supplied toilets. I ask him why Sulabh has succeeded where the state has failed. 'Government property is everyone's property. Stall doors are taken away overnight. People do not respect it.' They have problems in Sulabh toilets too. In the Indian Express Sulabh, Mohan strides as usual into the ladies and finds several women doing their washing on the floor while a tap in the sink gushes uselessly. He chastises them. They are pavement dwellers, he tells me. They don't know the meaning of taps or the value of water, because they've never had it before. The pavement dwellers are here because Sulabh gives them a free weekly pass. The women crowd around me outside to show me their ID cards. Name, occupation, the Sulabh logo of a woman carrying a headload of excreta, with a red cross through the image, residence.

Residence? 'I live near the bus shelter,' says one, and she means on the ground. 'I live on that pavement over there,' says another. They are officially BPLers – Below the Poverty Line – and are entitled to rations. Their piece of pavement is usually rented from a local slumlord or gangster. At least with the Sulabh card, they can wash for free. They can get some dignity along with rice.

Sulabh has innovated in other ways: some Sulabh toilets also house primary schools. Others have health clinics attached. As impressive an achievement as that is – as any traveller stuck waiting for an Indian train will appreciate – Pathak is proudest of the effects his business has had on scavengers. Sixty thousand have been 'liberated' as a result of Pathak's efforts. Some are given alternative employment as cleaners in Sulabh toilet blocks. Nonetheless, a Sulabh employee tells me that hierarchies still persist. Scavengers will always be the lowliest cleaners. 'The caretakers will be Brahmins, because they're the ones collecting the money.'

Six thousand wards of scavengers have also been given education at Sulabh Public School, whose premises are on the Delhi campus. It provides education with a message: the intake is 60/40 scavenger and non-scavenger, to encourage mixing. All children learn in English because it's the language of the educated, employable Indian. Classes are also taught in Sanskrit, the school headteacher tells me, 'because that is the language of the Brahmins'. Sanskrit is a shocking thing to teach a scavenger child, and therefore makes a powerful point.

Pathak tells me with pride about a programme in Rajasthan whereby twenty-eight scavenger women were given the opportunity to sell snacks for a living. It sounds a modest achievement. But even after three decades of trying to eradicate untouchability, getting an Indian to eat food prepared by a scavenger is a big deal. The Rajasthani snack women may be few, but the number is incidental. The point is that Indians will now buy food from people who used to disgust them so much, they would not touch their shadows. Even Pathak's Sulabh colleague Mulkh Raj admits that the habit of untouchability is still formidable enough to affect his own thinking. 'You can share lunch with scavengers,' he says, 'but you'd never marry a scavenger girl. It's wrong to pretend this doesn't exist.'

Pathak's other great achievement, in his eyes, was to make toilets talkable. 'In India until a few years ago,' he wrote in 2004, 'nobody could imagine that any politician, bureaucrat or businessman of some standing would like to associate his name with anything even remotely [...] connected to something as ordinary as a toilet. Now things have changed and most prominent politicians (including Ministers and Chief Ministers), high position bureaucrats and well-known businessmen readily agree and rarely decline to inaugurate the opening of public toilets in the country.' When Sulabh set up an adopt-a-scavenger system, top politicians invited scavengers into their home to share food, and sponsored their education.

For Pathak, the toilet was always a means to achieve his end. He said in one interview that 'Gandhi used the spinning wheel to enter families' homes; we're entering through the toilet.' Even his critics – of whom I find very few, despite looking – admit that he has changed Indian society. Paromita Vohra featured Pathak in her documentary about public toilet provision in Mumbai, *Q2P*. The film manages to be ethereal and earthy and features a fine example of Indian humour when, during a Take Back the Night march, some young women looking for a public toilet ask another marcher whether they can pee on a certain patch of ground. 'Can we sit here?' they ask. 'Is it a religious place?' 'Go!' says the woman. 'You sit and make it religious!'

Vohra has mixed feelings about Sulabh and its 'Visionary Founder'. 'If the goal of Pathak is to have eliminated scavenging, then he's failed. And I find Sulabh very paternalistic. But the guy is a Brahmin and he built the first public toilet in India. You can't take that away from him.'

Nor can anyone challenge his status as founder of the world's best-known toilet museum. In 1994, Pathak realized that maybe not everybody shared his delight in pour-flush privies and the transformation of scavengers to snack-sellers and cleaners. He decided to 'make toilets interesting'. During a visit to London's Madame Tussauds, he got an idea. Why not build a museum of toilets? Letters were dispatched to all foreign embassies in Delhi, asking for information about their country's toilet habits. The British provided a small booklet on the work of Mr Thomas Crapper. The Counsellor for Scientific and Technological Affairs at the US embassy could offer only the address of the American Society of Sanitary Engineering, and the suggestion that Sulabh's idea of playing the national anthem of various nations as one approaches their toilet in the exhibit might be 'something that many people might object to. A simple sign explaining the exhibit may be less controversial.'

Pathak gathered the exhibits on his global travels. The collection is impressive, but it still fits easily into a single room on the campus, next door to the biogas research lab. Replicas of historically relevant commodes, toilets and latrines are placed alongside a microwave toilet – used in ships – and a portable privy aimed at campers. There is a French commode disguised as English books, including a Shakespeare play. 'The French always used English titles for books,' says my guide, as if he can't imagine why. In the centre of the room, in a glass display case, there is a model of the Sulabh public toilet at Shirdi, supposedly the largest in the world, which has 120 toilets, 108 bathing cubicles, 28 special toilets, 6 dressing rooms and 5000 lockers, as well as a biogas system. (I later make a ten-hour round trip to spend an hour at Shirdi, and on the road wonder as usual where I will find a toilet, before I remember where I'm going.)

On the walls hang densely detailed displays relating to sanitary history. Visitors who trek out to the airport area for an education – they are more numerous, since the museum was included in the *Lonely Planet Guide to India* a couple of years ago – will get an education. They can learn that the best and first flush toilets were built five millennia ago in the Indus valley cities of Mohenjo-daro and Harappa, and that Ben Affleck once bought Jennifer Lopez a jewel-encrusted toilet seat (though I hear she's now moved on, to a TOTO Neorest). They can be enlightened by one poster that elucidates the Su-jok therapy devised by Korean scientist Park Jae Woo, which I will include here in a spirit of public health because it served me well during ensuing months of research in toilet-deficient places. Should the urge to defecate strike, take a pen, pencil or blunt object and trace a line, deeply and with pressure, in a clockwise direction on the left palm or anti-clockwise on the right. The urge, assures Dr Park, 'will immediately cease. You too can try sometime and feel the magic pressure in reverse order will give good relief in constipation.'

A visitors' book collects comments, some with expected humour, some serious. Jack Sim of the WTO has left his compliments. Nana Ziesche thought it 'such a big history part never taught in school. What a pity.' Swiss tourist Jonathan Hecker offered his congratulations because 'this is exactly what we need to pull sanitation out of its dirty corner'.

Pathak intends to keep pulling. He has plans for a University of Sanitation. He will also amend the non-profit model and accept grants. There is still much work to do and Sulabh needs help to do it. Despite the organization's achievements, half a million Indians are still cleaning dry latrines. 'Seen in that context,' Pathak tells me, 'Sulabh has achieved almost nothing.' A Sulabh colleague is also gloomy. 'Sanitation is a gigantic problem,' he says. 'The world needs a thousand Sulabhs, a hundred Dr Pathaks. What Sulabh does is a drop in the ocean.'

Pathak prefers to see things more brightly. 'We are still at the beginning of the beginnings,' he once said. 'We are a candle in the dark.' And the dark doesn't frighten him. This is the man who transformed teenage rebellion into a toilet revolution, and overturned profoundly held beliefs about purity and pollution in the process. 'It's totally amazing,' he tells me by way of a farewell. 'Scavengers used to be afraid of our shadows. But look. The earth and sky can meet.'

Da Li, China – *Author*

5. CHINA'S BIOGAS BOOM

A pig in every bedroom

A several-thousand-mile road trip across China can provoke questions. For example, how tiny dots on the map turn out to be cities that I've never heard of, but which have six million people and more skyscrapers than London. Or why Red, our Chinese translator, thought that asking melon-sellers for directions was always the best option when lost. Or why we were almost killed by truck drivers several times a day, occasions which provoked comments from our driver, a chubby, lovely man called Wang, like, 'That truck's from Inner Mongolia. He thinks he's still driving on grassland.'

Mr Wang had found my interest in *fen*, the Mandarin word for excrement, peculiar. Nonetheless, he tried to be helpful. He would point out when he spotted a truck full of *fen* looming behind, though its odour preceded it by far. He would alert me when he saw a tiny figure in a roadside field bearing a tank and a hose, spraying – by the smell of it – the contents of his toilets on his cabbages. This practice would horrify any public health professional, given the disease-load of faeces, but it's what happens to ninety per cent of China's excrement, and has been done for ever. There are reasons not to eat salads in China, and why the sizzling woks are so sizzling.

Of all the peoples of the world, the Chinese are probably the most at home with their excrement. They know its value. Those roadside *fen*-spreaders are only the latest practitioners of a 4000-year tradition of using human faeces to fertilize fields. China's use of nightsoil, as they rightly call a manure that is picked up at night, is probably the reason that its fields and paddies are still healthy after four millennia of intensive agriculture, while other great civilizations – the Maya, for one – foundered when their soils turned to dust.

Sanitation professionals sometimes divide the world into faecal-phobic and faecal-philiac cultures. India is the former (though only when the dung is not from cows); China is definitely and blithely the latter. Nor is the place of excrement confined to the fields. *Fen* and toilets have featured prominently in Chinese public life and literature for at least a thousand years. In Beijing, I found several shelves of books – under the heading Toilet Culture – at a state bookstore. One tells the tale of Qi Furen, concubine of the first Han Emperor, whose fate was to have her eyes burned out, her limbs cut off and her ears sliced off by her mother-in-law, who then threw her into the toilet to die. The dowager empress called the poor concubine 'the Human Pig', a name most records attribute to the total abasement of Qi Furen, but which I suspect had more to do with the fact that up to today, pigsties and toilets in China are often the same thing. According to some authors, the Chinese word for toilet, *ce*, originally meant pigsty. One of China's toilet goddesses – there are several – was originally a beautiful lady called Zhi Yan who saved a boy who was drowning in a toilet. The boy survived; Zhi Yan was drowned and made a toilet goddess by the King of Heaven. For reasons that remain obscure to both me and Red, even after reading the riches provided by the Toilet Culture shelf, 'the toilet deity is therefore a pig or a beautiful woman'.

In the Communist era, excrement took on political importance. Andrew Morris, a historian at California Polytechnic, relates the story of Chen Qiaozhu, a famous nightsoil collector from Shanghai who eventually managed to leave her sanitary profession, or – in her patriotic description – 'I came out of the toilet to declare victory.' (She also became a competitive cyclist.) In 1959, the nightsoil carrier Shi Chuanxiang was a star speaker at the Party's National Conference of Heroes. He vividly described working for the exploitative gangs who controlled Beijing's nightsoil collection, and of customers who showed their appreciation for his work by calling him 'Mr Shitman' or 'Stinky shit egg'.

Shi qualified as a hero because Party policy-makers had decided that excrement was essential for the Great Agricultural Leap Forward. States competed to collect the most valuable fertilizer. Hunan province launched the Seas of Shit, Mountains of Fertilizer Campaign, exhorting all Hunanese to collect as much human nightsoil as possible (they obliged with 10 billion kilograms' worth). In Hubei, 'ten thousand people entered battle like flying horses to collect manure and march forward side by side'.

Flying horses have now evolved into cars, but the Communist Party's efforts to dictate the toilet habits of its people are unchanged in their fervour. Since the 1930s, China's authorities have thrown much energy into biogas. Along with all the other stunning statistics China can provide, it can also claim to be the world leader in making energy from human excrement.

Biogas can be produced from the fermentation of any organic material, from wood to vegetables to human excreta. In an oxygen-free digester, which acts somewhat like a human stomach, micro-organisms break down the material into sugar and acids, which then become gas. Mostly methane, with carbon dioxide and a little hydrogen sulphide, biogas can be used as fuel for cooking hobs, lights and, sometimes, showers. It can also be converted into electricity. The slurry that remains from the digestion process is good fertilizer, and considerably safer than raw excrement.

At last count, if official figures are reliable, 15.4 million rural households in China are connecting their toilets to a biogas digester, switching on their stoves a few hours later and cooking with the proceeds. Worldwide, Nepal has more digesters per capita than China. India, meanwhile, has installed several million, though they run on cow dung, and there are only so many cows. China has a billion humans, and that means a billion suppliers of a cheap and inexhaustible supply of clean energy.

The scientific epicentre of China's biogas programme is the Institute of Biogas, or BIOMA, a complex near the American consulate in the city of Chengdu, in China's Sichuan province. The city is known for its panda research institute, one of whose residents is a sponsored panda called Microsoft, and is situated on a plain known as *Tianfuzhi guo*, usually translated as 'the Land of Abundance'. The lands around Chengdu are China's breadbasket. Here, farmers can get rich by the power of the soil. The capital of the farming lands is a spacious, pleasant city that contrasts with the smoggy, sweltering soup of Chongqing, a grim city of 10 million people that I had flown in to.

Mr Fang, a neat and pleasant man, is the deputy director of China's biogas programme, though he's only been in post for two months. Before biogas, he worked on agricultural mechanics at a rice research institute. He knows more about tractors than digesters. Perhaps for back-up, he has gathered a welcoming committee of half a dozen scientists, including a fierce woman who is clearly there on Communist Party business. Mr Fang has prepared a PowerPoint introduction to his institute's work. On-screen, China's biogas programme is going full steam ahead. Between 2001 and 2004, Guangxi province installed 250,000 digesters per year in rural households, the most of any province. Biogas is most popular in the southern provinces of China, where the climate is warmer (biogas digesters work better in temperatures above 10 degrees Celsius). To combat northern temperatures, though, two digester systems have been developed. The pig-toilet-vegetable system links a pigsty and toilet to the fields, whereas the four-in-one model adds the element of a greenhouse, under which the digester is installed. The four-in-one was developed in Liaoning province, where winter temperatures reach –30 degrees Celsius, and where a digester without a greenhouse to shelter it would only work for five months a year.

The Institute of Biogas was set up in 1981 with support from the United Nations Development Programme (UNDP). The building dates from then, and from a tour of the laboratories below – which house dull jars of brown test material in spartan labs and rickety-looking equipment – it shows its age. Nonetheless, according to Mr Fang, BIOMA is the only such research centre in China, and famous internationally. The Institute has already assisted Romania, Guinea-Bissau and Rwanda in setting up biogas projects. (I forget to ask whether it was involved in Rwanda's extraordinary biogas project, whereby half of the country's prisons generate gas and electricity from the excreta of genocidal murderers.)

The BIOMA building, says Mr Fang with some pride, also houses the editorial headquarters of *China Biogas* journal, and has completed more than two hundred research projects for the Chinese government. I should not be fooled by the age of the equipment (though Mr Fang also asks me to put in a good word with UNDP about more funding). Serious work is done here. Professor Zhang, one of the scientists present, is working on 'increasing gas yield by improving officially sanctioned pig diet'. Dr Hu is interested in microbiology and the fermentation process. Elsewhere, some of the institute's eighty-seven staff are attempting to reduce the risk of bird flu transmission. This is a real risk, since digesters require the close proximity of animal and human.

Despite the complexity of the research, the scientists insist that the system is simple. 'The farmer doesn't have to do anything,' says Dr Hu. There is no stirring, no caretaking. 'All he has to do is ensure there's input and use the output.' To translate: all he has to do is use the toilet, and then the gas. If I want further proof, they suggest I go to see it in action, at the nearby village of Mian Zhu.

If towns with six-lane highways running through them are villages, then Mian Zhu is a village. We are directed to stop by the side of the road and wait for an escort, which is revealed to be two officials from the Rural Energy Office. I stupidly never took their names, so I will refer to them as White Shirt and Red Shirt, with apologies for disrespect. They take us through a field by the roadside to a compound of three houses, each with huge wooden gates on which colourful warriors and symbols have been painted. On the other side of the warriors, a trail of blood leads to – or from – the pigsty. Mrs Tien, whose house this is says they've just killed a pig. They're now a six-pig family when ten minutes ago they were seven. She is smiling and excited. 'She's saying, "They've come to see our biogas digester!"' says Red, though of course there is nothing to see but a concrete circle in the ground. Digesters have to be sealed to work.

In Mrs Tien's kitchen, I am given what will be the first of many biogas demonstrations. They generally proceed like this: switch on the biogas. Sniff. Express pleasure that it smells only faintly sulphuric. Light the gas. Express more pleasure. Switch on the light. Take a picture (which never comes out well because the light isn't actually that strong). Mrs Tien seems genuinely excited to demonstrate her free gas, though I get the sense she's done it before. Red Shirt says that Mian Zhu is the centrepiece of the Ecological Homeland programme, and sixty per cent of houses in this area have digesters, making a total of 10,000 households. People like biogas, says White Shirt, and anyway it has an illustrious history. 'In 1957

Chairman Mao went to Hubei province and saw a biogas digester and said, "This must be well-promoted."' And it was so.

Sitting on an expensive sofa next to a huge TV screen in Mrs Tien's living room, I ask the officials if biogas has any disadvantages. They say there are none. Any problems can be dealt with by the technicians of the Biogas Management Unit, whose telephone number, since I ask, is 65980212. 'Each town has one to three teams of workmen, so they usually come within a day.' By 2010, they expect everyone in this province to have installed a biogas digester. If there are malfunctions, the network of rural energy technicians can fix them swiftly. It all sounds too good to be true. But, as an Indian biogas expert tells me, China can do this, because China can impose from on high. 'India is a democracy,' the expert says. 'We have to ask, to plead, to persuade. It takes longer. It is harder. China can do things faster.' Also, China can pay. Households that buy a digester get a 1200 yuan grant towards the total cost (usually about 3000 yuan). It all sounds perfect. Even so, I am not persuaded by the Potemkin village of Mian Zhu.

To reach the regional headquarters of the Chinese Women's Federation, which are located in the lovely city of Xi'an, one must first walk the length of a street of shops and barbers. The barber shops have barbering equipment and the added, odd extra of over-made-up women smiling insistently at my companion, a Caucasian man. At his approach, doors open and garish faces pop out before they see me and the smile disappears. After that gauntlet, I arrive at the Women's Federation in a sour mood, but it doesn't last, because my appointment is with Wang Ming Ying, founder and president of the Shaanxi Mothers' Environmental Protection Volunteer Association. Wang's energy can soothe the foulest tempers, even mine.

Her office is not much bigger than a cubicle, with a bed for the times she regularly works too late to get home. Wang is also small, but her spirit is disproportionate to her stature. She radiates warmth, even in translation, and she will not hear of starting a conversation without first providing platefuls of huge peaches and gigantic watermelons. 'You have to eat melons to help the farmers,' she says. 'This year there are too many melons. Eat! Eat!'

I'd read about the Shaanxi Mothers a few months earlier, when they received a sizeable international prize from a sustainable energy foundation. The group won for installing biogas digesters using human excrement in provincial areas of Shaanxi, where rural life is generally hard and the winters cold, and where any extra energy source makes sense.

In theory, biogas has all sorts of advantages and no disadvantages. It saves on artificial fertilizer use, because the slurry is filled with nutrients. Scientists from China's Research Institute of Medical Military Scientists decided biogas slurry increased vegetable yields by fifty to sixty per cent. One person's urine and faeces can fertilize 270 square metres of land, according to another calculation.

Because biogas can be used for cooking and lighting, it saves on conventional energy consumption of liquid petroleum gas (LPG) or coal (one cubic metre of biogas is the equivalent of six hours' worth of a 60–100 watt bulb). It saves forests, because wood is the commonest fuel source in rural China. A 1991 survey found that a five-person family in Guangxi used 2100 kilograms of wood for fuel per year. And it saves on labour: on average, a rural Chinese woman preparing food on an iron stove fed with rice stalks or wood takes two hours to cook a meal. A meal cooked from biogas can be ready in twenty minutes. Fast food.

Biogas has a long history, though the length of the history and its exact birthplace are debated. The Babylonians are credited with noticing that gas coming from sewage could be useful, as are the Assyrians. Marco Polo reportedly saw covered tanks of sewage – simple anaerobic digesters – in China. And Count Alessandro Volta, as well as inventing the electric battery and giving his name to voltage, concluded in 1776 that there was a correlation between decaying organic substances and the amount of gas they gave off. The Chinese say that they have been using biogas for as long as they've been spreading *fen* on their fields. But in biogas circles, the credit for pioneering the first noteworthy biogas digester is given to a leper colony in Bombay, which set up a digester using *gobar* (cow manure) in 1859 and gave Indian lepers the new concept of *gobargas*.

A discussion at the Institute of Biogas had produced the consensus that Chinese biogas owes its existence to a man called Luo Guo Ri, who developed a biogas digester in the 1930s and installed it in several locations around Shanghai. But both Luo Guo Ri and his plans were swallowed up by the Japanese when they invaded China in 1937. The plants were destroyed and that, for several decades, was that. In 1972, though, an oil crisis caused certain careful scientists to revisit the work of Luo Guo Ri. China had energy needs but, in the words of Mr Fang, 'our door was closed. We could not ask for support from outside. We had to find other energy sources.' If oil was expensive, then why not tap a resource produced, for free, by – according to population figures back then – 850 million bowels? A campaign was launched with the slogan 'Biogas for every household'. For the next five years, up to two million digesters were installed each year, in rural households that had toilets and enough pigs. To function efficiently and productively, animal manure is required to boost the volume of excreta. A cow's dung can produce 500 litres of gas a year, while a human only produces 30. There aren't many cows in China, but there are pigs. China's energy problems would be alleviated by a partnership between pigs and humans.

Except it didn't work. By 1980, half of the shiny new digesters were broken. The problem was humans, technology and the fraught relationship between them. These early digesters were made of brick and often leaked. Also, the Chinese academic J.X. Hong, writing in

China Biogas in 1993, estimated that a digester needs 30 kilograms of faeces daily to function properly, along with 50 litres of urine and water. This is the equivalent of the produce of twelve people, an unlikely family size in a country where couples were permitted one child only. Farmers with fewer pigs added rice stalks to boost the digestion volume, and the stalks caused clogging.

There was little training in maintenance, and no network of repair centres to step into the knowledge breach. When the units broke, they stayed broke. So the biogas digesters, despite their potential benefits, were abandoned, and China's farmers went back to their smelly *mao kun* – a hole-in-the-ground latrine – and to spreading the hole's contents on the fields, crude and teeming with pathogens.

Once again, farmers' wives set off each day to cut wood from the forests. Deforestation reached a scale that alarmed even the far-off government in Beijing. Thoughts turned to the broken digesters, and to how they could be improved.

At the same time, other thoughts were occupied by the phosphorus question. Human excrement contains nitrogen and phosphorus, which is plant food. Plant food is worth money. This economic fact oiled the wheels of the nightsoil collection industry enough to make its operators powerful. By the late nineteenth century, the leaders of the nightsoil trade had been given the ironic title of *fenfa*, or shit lords. The nickname was derived from *junfa* (warlords) and neither were to be messed with. In 1925, when Beijing police told the *fenfa* to move their drying yards – where their product was aired, producing awful odours – they only had to go on strike for three days before public disapproval forced the police to back down.

In 1930s Shanghai, the place dedicated to the loading of nightsoil onto barges had acquired the name Golden Wharf. This was not ironic but a genuine appreciation of the value of the stinking cargo it carried away to China's farmers. A Shanghai gangster's moll ran the wharf, and a mighty business empire, for decades. Her real name was

Sister Ah Gui, but she was known to all as Fen Huang Hou, or Shit Queen. Whether or not she objected to her nickname is unknown, but records show that she made a fortune from operating hundreds of nightsoil carts. Her profits were $10,000–12,000 a month.

By the 1940s, when Shanghai had glamour, electricity and telephones, night-stools (wooden buckets used as indoor latrines) could still be seen lined up in an alley of an early morning. Nightsoil collectors still drove their wooden carts through the streets, always courteously informing housewives who might have overslept, 'Leaving! Won't be back!' before they left and didn't come back. The night-stool was still a standard part of a Chinese bride's dowry, and 'red-dyed eggs were placed inside the night-stool to serve as a symbol or wish for the birth of a child'. By the 1980s, a million Shanghai households still had a night-stool, or 'honey bucket'. As late as last year, a *New York Times* journalist living in Shanghai wrote of her neighbours in her alleyway bringing out their chamber pots to be collected, though the nightsoil man's wooden cart had been replaced by a municipal truck.

But the eggs in the night-stools worked too well. In the vicious circle of nightsoil fertilizer production, more people were produced and more people produced more nightsoil, which produced more crops, which fed more people. Eventually even the land of the *fenfa* had to resort to artificial fertilizers, which usually contained phosphorus. China again is agriculturally advantaged, having one of the largest phosphate reserves in the world, but even she cannot make the finite infinite. Estimates for when the world's exploitable phosphate reserves will be exhausted range from 60 years from now to 130, but they will run out sooner if current rates of exploitation continue. Biogas made agricultural, financial and energetic sense.

Perched on the bed in her Xi'an office, Wang Ming Ying tells me that for her, it was all about the trees. In the mid-1990s, she worked in the Propaganda Department of the Women's Federation. In 1995, she attended the UN women's conference in Beijing, and it changed her life. 'I saw,' she tells me, 'how the poverty of women is directly related to the deterioration of the environment.' Poor rural women try to clear more land for crops by cutting down forests. This brought on soil erosion, so more forest was cleared for new crop land. It was a vicious cycle that no-one knew how to escape.

Wang Ming Ying set off to northern Shaanxi province 'to see what was going on'. She found hillsides empty of trees and farmers devoid of hope. 'I thought that if a woman has education or not, we can do environmental protection together.' She decided to form an organization of women. Mothers, actually. 'Mothers are key: they can influence the family.' She named the group the Shaanxi Mothers Environmental Protection Volunteer Association – Shaanxi Mothers for short – and applied to the authorities for registration. They wouldn't give it.

The group's name was surprisingly controversial. 'They didn't like the word "volunteers".' Voluntary activity was a problematic concept in China then. Though the word appears in the Chinese national anthem, there was no tradition of volunteering, because public service was always imposed from above. The state controlled everything, and that included excreting habits and public hygiene. Throughout the 1950s, for example, the Chinese government tried several times to eradicate a plague of schistosomiasis, an infection of a parasitic worm found in dirty rice paddy water (it is also known as bilharzia or, in Chinese, 'blood-sucking worm disease'). Shepherd boys, according to a report, 'were mobilized to pick up stray excreta'. Elsewhere, there were campaigns such as this one in Tientsin: 'Street cadres, together with the chief of the local agricultural producers' co-operative, members of the residents' committee, and the farm officer in charge of fertilizer accumulation were asked to adopt the sanitation methods used in model villages. Mass meetings were held in the village, and methods of garbage and excreta disposal, treatment of polluted water, cleaning of pig pens, sealing of manure pits, etc., were demonstrated. Finally, whole-hearted participation of the entire village was achieved, and the problem of sanitation was solved.' Simple.

But Wang Ming Ying persisted in wanting to be a true volunteer. 'For the longest time, everything we did was under command. Now we wanted to do something not because someone was paying us or ordering us but because of our own initiative.'

After a few years of environmental work – there was, for example, a lot of litter-collection – Shaanxi Mothers were shown a video of biogas technology. They liked it, and decided to try it out with two test families in northern Shaanxi. The families lived in a village that had a fate typical to the area. Thirty years earlier, its population had consisted of four families, and the village was surrounded by trees. By the time Shaanxi Mothers arrived, there were thirty-four families and the forest was almost gone. 'Their way of surviving was wrong,' says Wang Ming Ying. 'They were cutting more wood for fuel but cutting down their livelihood.'

Biogas was an ideal solution. But when the Mothers arrived for a follow-up survey, neither digester was being used. They couldn't understand it. 'Everybody was so hostile. But biogas was such a good concept!'

The answer lay in the interaction of humans and technology. For any technology to succeed, everyone – especially the crucial early adopters – must fully understand how to operate it. Something had gone wrong with the education process. For months on end, no-one in the village would explain why they didn't want biogas. Then one day Wang Ming Ying was talking to Qiao Liu Ye, the mother in one test family, when Qiao mentioned, almost casually, that her young son had drowned in the digester. Wang Ming Ying was horrified. The family had left the lid off, and their toddler son Peng had wandered up to the hole and fallen in. Whether the tragedy arose from poor instruction or poor comprehension of the instruction was unclear. But Wang Ming Ying learned a lesson. You can't install technology (the hardware) without ensuring that the human element (the software) is also operational. Follow-up is essential. Shaanxi Mothers set about soothing the trauma of Peng's death by the simple method of talking, a lot. It took months, but it worked.

Ten years on, Shaanxi Mothers have installed 1294 digesters in 26 villages. They have won prizes and funding, though never enough. The money goes to subsidising a third of the cost of a digester, with the householder and the government making up the rest. Wang Ming Ying estimates that for every new biogas digester installed, 1.2 tonnes of firewood – three trees – will be spared.

Wang Ming Ying unrolls some large hand-painted posters that are used to instruct villagers how to operate their digesters. The advice runs from the obvious (keep the lid on and small children away; don't have your lamp too close to the ceiling; don't install the stove in your bedroom) to the more complicated (stir the slurry periodically to add oxygen to the mix, which feeds hungry micro-organisms and helps break down the organic matter). Every so often, mix in some animal dung, which adds more valuable bacteria. Sit back and wait for the benefits. They are many. The excrement is safely contained in the digester, so 'the pigs and sheep don't stink like they used to,' says Wang Ming Ying. The fly population – significant, with all those pigs – is reduced by sixty-four per cent, on average. Most importantly for the Mothers' target population, the backbreaking job of chopping firewood has been cut down. Women now have an unusual thing called free time.

The journey to Da Li is long. It goes along roads that are so new, they're not on the map and roads so bad, they are flattened rocks with aspirations to being a thoroughfare. After several hours of bone-rattling driving, we arrive in north-west Shaanxi province, to a village of apples. There are boxes of apples everywhere; being loaded onto trucks; stacked on street corners. Later, I see the same brand for sale in Hong Kong. This is apple country. What the buyers of apples probably don't know is that this is apples-fertilized-with-human-excreta country.

For once, I have not planned well for bathroom breaks. This is unusual, because any woman who travels in India, China or a country relatively devoid of public toilets learns two things: go before you go and always wear a skirt, the better for squatting. I am wearing a skirt, but this is a busy village, and I won't use a public roadside. By the time we reach the meeting room in the village council offices, I am desperate.

'The toilet? Yes, we have a toilet,' says one of our hosts, a friendly woman wearing an enormous sun visor on a dull day. She leads me outside, along the street, over some roadworks and to the only public toilet in town. There are four stalls – if holes in the ground separated by bricks count as stalls – no doors and many maggots. I can be composed about travel trials, but this makes me furious. The roadside would have been cleaner. It is unfit for humans, yet better than the facilities that most humans have.

In the meeting room, there is a delegation of half a dozen villagers. We have been sent by Wang Ming Ying, who is a hero here, and all due courtesy is being extended. A blackboard bears the phrase, 'We wholeheartedly welcome the advice and arrival of our superior leaders', and bowls of apples and grapes have been thoughtfully set out on the table. They have been fertilized with biogas slurry, the village leader tells me with pride. 'Look,' he says, 'how juicy the apples are. They are better now that we use biogas. The skin is thinner and the juice is sweeter. Even rice is better: rice cooked with biogas is chewier and less likely to stick.'

The sun visor woman reads from a sheet of statistics about the wonders that biogas has brought to the village: 1506 people, 368 households, 790 women, 378 working women, 104 digesters. She says that biogas digesters have reduced three things. 'Chemical fertilizer, agricultural insecticides and women's household labour. There have been three changes: human and national excreta is now turned into treasure. Households are much cleaner. Neighbours have a better relationship.'

Also, farmers' incomes have increased. Annually, they save 1400 yuan (US$200) on fertilizer, fuel and the medicines they would otherwise have to buy for the constant diarrhoea and stomach illnesses caused by filthy latrines. She moves from diarrhoea to energy. You save two canisters of cooking gas per year, worth 120 yuan ($20). Using biogas for lighting saves another 40 yuan ($5) on energy bills. All in all, she concludes, the village has increased its income by 300,000 yuan ($43,000) a year. 'The village,' she concludes firmly, 'is happier and wealthier.'

The Shaanxi Mothers arrived in Da Li in 2003. 'Everyone here knows Wang Ming Ying,' says the village leader, a jolly man called Zhou. 'She comes every month to do checking and follow-up. Everyone loves her.' Next to the meeting table is a loom, because the Shaanxi Mothers didn't only bring biogas. 'They came to liberate women,' says Mr Zhou. 'Now they have time to do weaving, to earn a bit more money.'

Mr Zhou speaks in a strong Shaanxi dialect that the translator finds tricky, and has a farmer's tan and neglected teeth. He was the first villager to install a digester, though it was part of a government biogas campaign that started before the Shaanxi Mothers arrived. He says it wasn't a difficult decision to get one. 'I have a lot of pigs and they create a lot of excrement.'

He doesn't mention human manure until I ask. 'That was also a problem. Before, we would cover our shit in mud in a hole at the back of the house. When we needed it on the fields, we'd bring it over. But sometimes children would play with it. They might destroy the pile and scatter it around the courtyard.' Such a hole is called a *mao kun*, meaning 'straw hole'. 'It's not a hole made of straw. *Mao* is wild grass, unwanted grass. So *mao kun* is a hole that no-one wants. It is something completely undesirable.' Villagers regularly had worm infections or dysentery, but they thought that they were stuck with it. They thought they couldn't afford anything else.

Studies show that people with no latrine or a poor latrine regularly say cost is the biggest obstacle to improving their sanitation, even though a decent latrine pays its way in health and economic benefits. In fact, though the Da Li villagers were dissatisfied, it wasn't enough to spur them to change. They carried on with the *mao kun*, even in the awful bitter winters of northern Shaanxi, when the ground freezes and the temperatures reach –11 Celsius.

In Da Li, as in countless other villages, things began to change when the city came back to the country. A hundred and twenty-six villagers had left to work in the city over the previous three years. They got used to different standards of living. 'Young people were coming home and complaining about the *mao kun*,' says Zhou. 'They didn't want to use it any more. They couldn't deal with it.' They demanded better facilities for their visits home. The women of Da Li proved to be powerful allies. The reason why becomes obvious when Mr Zhou leads me to his house and into the kitchen, past the cartfull of apples in the driveway. Here, his wife gives me a demonstration of how she used to live and breathe. She kneels in front of her cast-iron oven, pretending to feed it with kindling and rice-stalks, and mimes how she used to cough and how her eyes would water. The ovens are still used to bake bread, but otherwise the two-ring biogas burner is enough for three meals a day in summer and two in winter.

Mrs Zhou's saviour lies underground. She shows me the square of concrete it lies beneath, and the cleanish latrine behind a curtain next to the pigsty, so that pig and human excrement can be easily sluiced into the digester. Mrs Zhou uses the time she once spent cooking to work more in the fields. Other biogas families have diversified. In one house nearby, in a small courtyard behind a gorgeous wooden carved door, a Mrs Yang sits weaving at a loom as if she's done it for centuries. But she began only a year ago, when the Shaanxi Mothers introduced a new way to fill the hours no longer spent choking on fumes by an oven.

Biogas is not perfect. As the tragedy of Peng showed, digesters can fail because of mechanics and human error. Also, there is little agreement on how safe the slurry actually is. Opinions vary as to whether a four-week digestion process, for example, kills all pathogens. Ascaris eggs, which grow into long and revolting worms, are exceptionally hardy. (They are also still unvanquished, though humanity has been dealing with them for ever: ascaris have been detected in fossilized Peruvian dung dating from 2277 BCE.) Swedish academic Mathias Gustavsson, a fan of biogas – he refers to it as a 'solution in search of its problem' – writes that 'there is no such thing as a total removal of all parasites due to an anaerobic process'. But a biogas digester has to be better than a bucket.

There are other quibbles. Biogas lamps get very hot and are a fire risk if they're suspended too close to the ceiling. The Rural Energy officials of Mian Zhu tell their customers to light a match and wave it around before they light the gas, because if there's more than five per cent methane in the air they might get dizzy.

Also, the impressive numbers handed out by the Institute of Biogas should be taken with circumspection. Jiang Ping Zhao, a Senior Energy Specialist with the World Bank in Beijing, believes that, technologically, biogas has evolved far enough to be easy to use. The days are past when state officials would install a digester, then be nowhere to be found when it broke down. Yet the numbers give pause. Chinese government targets project that 80 million digesters will be in place by 2020. They will produce 40 million cubic metres of gas. Jiang thinks this is odd. 'They don't have meters on the digesters I've seen, so where are the figures coming from?'

There are also questions about the usefulness of biogas outside pig-owning rural households. Since 1980, China's urban population has more than doubled. By 2025, over two-thirds of Chinese will be city-dwellers. Their energy needs will be immense. But the current design of household digesters is little use in urban areas, where livestock is absent. One survey of the use of biomass and coal fuel dismissed biogas digesters by saying that 'they are limited to areas with sufficient dung, water, temperature and financial capital'. I make this point to the Rural Energy Officials at Mian Zhu, as we drink endless cups of green tea in their ornate meeting room. China used to have the famous Communist slogan of 'a chicken in every pot', so why not rework it for biogas? Why not get apartment owners to keep a pig or two? Yes, they say. 'A pig in every bedroom!'

We laugh, but it's a serious question. Advocates of biogas promote it as an impressive energy source that need not be confined to village backyards. There are already some large-scale biogas digesters in Germany and Sweden, but more interesting are the buses of Lille. In 1996, the northern French city began to convert its bus fleet to run on biomethane, a fuel derived from municipal sewage (other sources are organic kitchen waste). Ten buses now run on biomethane and ninety more on natural gas. City officials swear biomethane is cleaner. They demonstrate this for reporters by sticking into an exhaust pipe a white handkerchief that emerges the same colour. Biogas emits fewer particles and twenty per cent less carbon dioxide than conventional fossil fuels. Also, it is economical. Infrastructure is expensive, but because the gas is cheaper than diesel or petrol – and its raw materials are supplied for free – it matches petrol in price. What is more, the Lille officials say, it doesn't damage food sources. (The practice of farmers dedicating their land to grow crops to produce bioethanol was condemned by a senior UN official as 'a crime against humanity'.) Of course there are obstacles to the brave new world of biogas: infrastructure is sparse. The oil industry is a powerful disincentive to alternative fuels. But biogas deserves a bigger place in our future, because of how it has so far transformed the present.

In Da Li, I make a last stop at the house of another Mrs Zhou, a white-haired widow who lives with her pigs and her beloved digester. She thrusts a bag of plums into my hands. She offers apples. But I decline. If I won't take the fruit, she insists, then would I like to hear a poem? She has written it in honour of Wang Ming Ying, whom she idolizes for having made her life easier. 'I'm illiterate,' she tells me, 'but I memorized the words.' I watch her standing there in her yard, ramrod straight, equipped with time, health and pride, and all because of methane and excrement. She starts to recite:

Hear me out, friends!
Let me tell you about biogas
Our Chairwoman is a vanguard of the environment
She knows biogas makes good harvest
Government cadres listen to her with delight
Quickly they build the biogas tank
Cooking is easy with biogas
Relieves us women with big problem
Four generations of the family can now dine together
Soup noodles for breakfast
Dry noodles for lunch
Biogas, what a blessing!
The elderly now stay home

The birth of piglets can be managed
Everyone is happy
Everyone wants biogas
Except that money doesn't come easy
We still need to work on it.

Public toilets, North Shields, England – *The Caravan Gallery*

6. A PUBLIC NECESSITY

Frightening the horses

At the Happiness and Prosperity service station in the rural reaches of Sichuan province, I prepare to face the public toilet. We have been driving for hours, and my need is pressing, but I hesitate because Red has told me what lies beyond the entrance. She grew up in flushed and plumbed Hong Kong, and now lives in flushed and plumbed Britain, and she has the ex-pat's snobbery about standards that were before familiar and are now primitive. So she won't come in, though she's desperate as well. It's not because the service station is unclean: the restaurant is pristine, and the food cheap and fabulous. It's because of the doors. There won't be any.

In China, this type of public convenience is called open-style or *ni hao* ('hello'). Open-style stalls are common enough to be unremarked upon in Chinese public life and to be mentioned in travel guides warning first-time visitors of Chinese uniqueness. The 2002 film *Public Toilet*, by the Chinese director Fruit Chan, demonstrated the place open-style toilets hold in Chinese culture:

banal, familiar, normal, where men squat and chat, unashamed and unabashed.

This is my first open-style experience. I ask Red if there is an etiquette. Where should I look? What is considered rude? Is it obligatory to say *ni hao*? I have no idea, because this is turning all my concepts of public and private upside down. I know that some schools and institutions in the Western world have doorless toilets, the better to foster compliance or – in the case of the military – to extract individuality. But I grew up in a culture that provided privacy abundantly and without question. I like doors. At the Happiness and Prosperity service station, I know I will miss them. Red shrugs, trying not to smile (and also, I hoped, trying hard not to pee).

There is a line inside. Women lean against a curving wall, only a few feet away from half a dozen women squatting in the stalls opposite, over squat latrines placed above a channel of trickling water. There isn't a door in sight. No-one says hello. I lean into the wall, making no eye contact and hoping to go unnoticed, but this is untouristed China, and I stand out anyway. The women in the queue smile at me. They gesture. You go first! No, please! Possibly it is courtesy. Probably it is curiosity. Let's see how the *lo wei* (foreigner) does it!

How does the *lo wei* do it? With my head down, the fastest expulsion of liquid possible, and by building doors in my mind.

It wasn't too difficult actually. If I thought about it, I'd had decades of practice. Privacy in the twenty-first century relies on hard, material items, on shielding wood or shiny partitions. But I've never found any that were truly soundproof, that provide what one earnest government report on school toilets called 'aural privacy'. Doors and locks are a convention that takes privacy only so far, as I discovered the day in high school when my friends decided to lean over from the next cubicle. Privacy in a public toilet relies on an assumption –

a social pact – that something that is not seen is not heard. Privacy involves pretence.

To perform a private function in a public setting requires what the American sociologist Erving Goffman called 'civil inattention', or the art of making living among unknown people tolerable. You know they're there, but you pretend not to acknowledge them, by whatever means are at your disposal. How many people have lingered in a cubicle so that the sound of their excretion – of whatever variety – can't be associated with them when they come out? How many have cringed in a hotel bathroom too close for comfort to a bedroom containing a new lover? I have, and I will.

The modern concept of privacy seems as fixed as those doors, but it is actually an historical upstart. As Norbert Elias wrote in *The Civilizing Process*, it's only slowly and without much forethought that humans in industrialized societies have developed the pressing need to perform certain functions away from the gaze of other people. This, he writes, shouldn't necessarily be seen as progression. Civilization is too fluid a concept to be pinned down as a linear evolution from A (barbaric Middle Age manners) to B (superior modern propriety). Many modern habits would horrify our courtly ancestors. Nose-blowing, for example. It is still frowned upon in Japan to discharge one's nose in public, but modern 'civilized' Westerners happily – and to me, puzzlingly – continue with a habit that the Italian poet Giovanni Della Casa found unpleasant several centuries ago, when he wrote that it was not seemly 'after wiping your nose, to spread out your handkerchief and peer into it as if pearls and rubies have fallen out of your head'.

The rise of privacy, along with technology that could flush away evidence of defecation, allowed society to turn a natural bodily function into a hidden, shameful one. Our courtly ancestors happily defecated and urinated in public and without embarrassment. Public toilets have always existed, but until 200 years ago, they were always communal. The Romans dined and bathed in company, and they did the same with defecation. Their communal latrines, called *forica*, were lavishly decorated with marble and had running water which served to cleanse the sponges on sticks they used to wipe themselves. Romans had murals of leaping stags and floors of geometric mosaic. Another famous public restroom, the twelfth-century Longhouse of London, had 64 seats for men and 64 for women. Facilities were basic: wooden seats were placed above holes that dropped directly into the Thames below. Stalls were judged to be unnecessary.

The transformation of defecation into an activity only done out of sight and smell arose from population growth and the social change that accompanied it. As cities grew more dense, private space became a privilege of the elite. Individual cleansing facilities were out of reach of the masses, unless you counted the enterprising gentlemen who tramped the streets of Vienna, Paris, London and Edinburgh, wearing large cloaks and carrying buckets. Passing citizens in need could use the bucket as a toilet and the cloak as a cover, or an early, cloak-like door. (The street-cry of these human public toilets was surprisingly subtle, in times so excrementitious that the streets were sewers: the slogan of French ambulant toilet providers was, 'Every man knows what he has to do and it costs two sous to do it!')

Privacy was something rich people had, until the needs of industrialists and sanitary reformers coincided in the early nineteenth century. The mass labour force needed to be kept healthy to be productive, and the new device of the flush toilet was deemed to be the solution. (Not everyone agreed. In 1857, the happily named Mr Tinkler took his local council to court after they forcibly installed a flush toilet in his cottage, when he had been content with a privy.) Work and health also intersected in the landmark 1848 Public Health Act. A more mobile labouring population which no longer worked in the fields – where any bush constituted a public convenience – needed facilities on the way to and from work. The Act, therefore, required the construction of Public Necessities 'to alleviate public stench and disease'.

These Public Necessities would be something new. For the first time, the privacy of residential facilities could be provided in a public setting. In 1851, over 800,000 people paid a penny – the spending of which gave English another appealing euphemism – to use private-public facilities installed by George Jennings at the Great Exhibition, and the template was fixed. So was the importance of public necessities in civilized life, at least for the next century. Providing publicly funded facilities was simple good manners – and kept the smelly working classes from being too smelly – and the Victorians liked to show theirs off. Their public conveniences were as lavish as those of the Romans, with pretty porcelain and copper piping, heavy wooden doors and sturdy locks. Over the Channel, the Parisians followed suit. A 1903 toilet, the first in a station of the Paris Métro's new Line 1, provided thirteen stalls for men and fourteen for women, three of which included bidets with warm water. There were six attendants, and the toilets were open from 7 a.m. until midnight.

From where I'm standing, that Paris subway toilet – long since closed, by the way – sounds like heaven. I grew up in the final decades of the twentieth century. This makes me a child of the new dark ages of the public toilet. In London, forty-seven per cent of public toilets have closed over the last eight years, and nationwide, public toilets have decreased in number by forty per cent. In a 2001 guide to New York City, Fodor's warned visitors that 'public bathrooms are few and far between, and run the gamut when it comes to cleanliness'.

Three successive mayors failed to agree to proposals that would have put public restrooms on New York's streets, and when Michael Bloomberg did, in 2005, newspaper headlines cheered that new restrooms were finally going to be provided. On closer inspection, it was apparent that the plan would provide 3300 bus shelters, 330 news kiosks and only 20 public toilets. Most of the world's cities lack public facilities, of course. A Russian professor, lecturing at the WTO's Moscow conference, managed to make the deprivation of Soviet times sound poetic. 'When we still had socialism,' he explained, 'the way to survive was to take your heart in your hand and squeeze tightly and be very, very patient.' It was the same with public bathrooms, which were few and very, very far between. 'Sometimes there are kilometres to run before you reach a facility and can unsqueeze your heart and do what your body wanted you to do.'

But New York City and London are supposed to represent the height of civilized achievement. They are examples of what urban planners call 'the sanitary city'. They are cities that are supposed to have already perfected the skill of delivering services to their citizens. Jawaharlal Nehru, India's first Prime Minister, declared that 'a country in which every citizen has access to a clean toilet has reached the pinnacle of progress'. In that case, progress in two of the world's most advanced cities is going backwards, with barely a protest. Academia doesn't much care either. Toilet culture in general, and public toilets in particular, have rarely been considered worthy topics of study. Outside Norbert Elias and Alexander Kira, there is little examination of public facilities. When a call for papers was put out for a journal to be entitled *Toilet Papers: The Gendered Construction of Public Toilets*, a blogger at the *New Criterion* magazine could not contain his scorn. This, he wrote, represented 'the pathetic intellectual and moral bankruptcy of the humanities. [...] Public toilets! Has it come to this?'

I find this strange. Anthropologists and sociologists should be infesting public toilets. There's nothing else in human society quite like them. Not in society, not quite out of it. Needed but rarely demanded. A place where all sorts of human needs and habits intersect: fear, disgust, conversation, grooming, sex. It's an ambiguous space that is not quite in the public eye, though the public uses it. A place of refuge and sociability; of necessity and criminality. How we are allowed to behave in toilets even influences everyday speech. Steven Pinker, in an exploration of taboo words, quotes a spectrum of excreta-related swearing. Shit is less acceptable than piss, which is less acceptable than fart. And so on through to snot and spit, 'which is not taboo at all. That's the same order as the acceptability of eliminating these substances from the body in public.'

To be uninterested in the public toilet is to be uninterested in life.

In the absence of academic curiosity, I will ask the experts. I have like most people used all sorts of public toilets in my lifetime. I have struggled with luggage through stupid turnstile toilets in railway stations; I have used clean and free toilets in supermarkets and libraries; I have trekked through half a dozen departments of department stores to find their poorly located, poorly sign-posted facilities behind indoor furnishing or haberdashery. I have sneaked into hotels, and failed, usually, to sneak into pubs, in case I'm identified as not being a customer and because I have an English sense of embarrassment. I have bought cups of coffee and bottles of water that I don't want, to be able to use café restrooms.

All these qualify as public conveniences, though the British Toilet Association prefers the term 'away-from-home toilets', which mistakenly presumes that all of us have homes, or that our home and the public toilet aren't the same thing. A postman in Devon was disconcerted to find an envelope addressed to 'Simon Norris, The Disabled Toilet, The Pleasure Gardens, Bournemouth'. My local government in Hackney was recently obliged to remodel the security systems of a brand new public convenience because it was being used as overnight accommodation by Polish migrant workers, undoubtedly delighted to get a smallish room for 20p a night. (Fighting over the more spacious disabled toilet was fierce.)

Where there have been no facilities available, I have done what ninety-five per cent of Britons have done, according to a survey by the Keep Britain Tidy Campaign: I have squatted behind parked cars, or in alleyways, with a friend keeping watch, or with considerable anxiety when I am friendless. Along the way, I have talked to toilet attendants in several countries, including in China, where a woman called Shu told me that not one of her clients flushed the toilet. She was also sick of finding footprints on the seats, from Chinese peasants too accustomed to squatting to change their habits for a new contraption.

In Bangkok, I met women drivers of mobile toilet trucks, steel and shiny and smelling of ammonia, who seemingly lived in their cabs, which had carpets and TVs and Buddhist shrines. These hardy women would step in with their dozen toilet stalls at public events, or at the airport, which lacked toilet facilities, or to assist Bangkok police with random drug testing that required impromptu urine samples. A woman who wouldn't give her name told me she had driven a mobile toilet truck for twenty-six years, and her parents did it before her and her daughter will do it after her. She says some of her neighbours give her a hard time. 'They say it stinks and that I stink too. But as jobs go, it's fine.'

At various points along my public necessity life, I used the Longcauseway facilities in my home town of Dewsbury, a medium-sized town in Yorkshire. The Longcauseway toilets were built in the 1980s as part of the Princess of Wales Shopping Centre, which replaced the old bus station with the kind of low-rise forgettable shops that have infected the minds of urban planners for decades. I had used them many times while growing up, and remembered their heavy doors with old-fashioned locks that could only be opened by the insertion of a coin (2p when I was young; 20p now). But mostly I remember them from a visit I made a couple of years ago, because at the time they also contained a sink full of soft toys and several jars of lemon curd for sale, by a sign saying, 'To pay, apply to Margaret'. When it came to writing about public necessities, of all the toilets in the world, I remembered the lemon-curd ones. I set out to find Margaret.

It takes a few visits, but I find her eventually, sitting in a cubby hole provided for attendants. It is furnished with two armchairs, a TV and VCR, books and tea- and coffee-making material. There is no shrine. It's a cosy place, and Margaret looks comfortable in her straight-backed easy chair, but despite the lived-in look, she says she's not here much. 'I'm always jumping up and down,' she says. She's always dealing with complaints, clients, cleaning. There's never a moment's peace.

Margaret has been working as a toilet attendant for three years. She doesn't like the job much – 'you'll not find anyone who does' – but she gets to raise money from her sink. (The soft toys and lemon curd are sold to generate funds for a cancer charity.) It's a perk of the job, and it makes up for the fact that, as she discovered recently, the council pays street-sweepers more than toilet attendants. Margaret is insulted by this hierarchy. When did a street-sweeper have to do everything she does? 'I've had to dress old ladies, take someone to hospital, look after children. This isn't just cleaning: I'm a care worker who works in a toilet.'

There have always been people whose jobs consisted of cleaning up other people's excreta. And because most societies have rules about what is clean and what is not, the people who have to deal with the unclean suffer the societal consequences of the time. However dismal Margaret's working conditions, though, she is still better off than the cesspool cleaners of the fifteenth century, who supplemented their income by charging fees to lance the boils of tuberculosis sufferers. In 1895, when Paris authorities removed the gas stoves of the city's famed *Dames-Pipi* ('Pee Ladies') toilet attendants to save on heating bills, the good women were forced to go on strike until their stoves were returned (and only because the pipes were freezing over).

Margaret isn't likely to resort to industrial action. Most people treat her with respect. They appreciate her, enough to buy her cakes and Christmas presents. But sometimes she doesn't appreciate them. Margaret has the lament of the toilet cleaner: you can't imagine how people behave. They do things they'd never do at home. In *The Bathroom*, Alexander Kira explains public convenience behaviour by noting that people feel more negatively about public toilets than private, because of the stranger factor. Public restrooms fuel old, primeval concerns about territoriality, which should be guarded, and strangers, who should be feared.

Kira lays out a spectrum of toilet tolerability. Most people are comfortable in a hotel toilet because it does the best pretence of being private. Workplace public toilets are the next best thing, because the people who use them are known, usually. And so on, through cinemas and shops and hotels, to the free-standing, filled-with-strangers public bathroom, which provides privacy from others, and a total removal of responsibility. People do all sorts in public toilets because they can, especially when prudishness persuades planners to locate them out of sight, away from public thoroughfares, behind hedges and in far-off car parks, where anything and anyone goes. Margaret sees all sorts in hers: Asian ladies in headscarves who sneak quick smokes in the cubicles and think she doesn't notice. Gin and beer bottles rolling out from under the cubicle doors; women who let their children eat food in the toilet. Drug tools. Things you wouldn't believe.

The removal of inhibition can be liberating as well as criminal. Recently, a Reuters reporter expressed frustration that American soldiers stationed in Iraq would tell him nothing, until he went to the latrines. 'You have to go out to the Port-O-Potties. For some reason, they talk there. You can read how they really feel – all the anti-Bush stuff, all the wanting to go home – in the writing on the shithouse walls.'

But generally, when local governments close toilets, they do so with the excuse that the facility is being used for drug-taking or sex. And sometimes it's true. The toilets in the small Yorkshire village where my parents now live were closed several years ago, when they were found to be listed on a website of the best places for gay cruising. When the House of Lords debated the ins and outs of sex in public toilets, the debate was colourful and sustained. Baroness Walmsley of Sutton Coldfield, though proclaiming herself a libertarian, agreed that a specific clause was necessary to prevent people engaging in activity that might 'frighten the horses'. Much aristocratic brainpower was expended on whether sex performed behind a closed cubicle door constituted a public or private activity. The peers made their point; the government backtracked, and the crime of having sex in a public toilet, even behind a closed cubicle door, was retained in Clause 71 of the Sexual Offences Act 2003. Transgressors can get six months in prison. In practice, hardly anyone does. It costs money to prosecute someone. It's cheaper to close the place down.

The nearest great city to London does things differently. The mayor of Paris, Bertrand Delanoë, known for interesting initiatives like rentable bicycles and river-side beaches, gives public bathrooms the attention that is their due. In 2006, he made all automatic public toilets – *les sanisettes* – free of charge. Usage grew from 2.4 million to 8 million visits in three months. He recently attacked public urination by fining transgressors 500 euros each. But fines and free toilets still didn't work: during the Rugby World Cup, the mayor was reportedly dismayed to see Parisians peeing against his town hall though there were sixty-two *sanisettes* nearby. His latest weapon is an 'anti-pipi' wall, whose angles spray the urine stream back onto the offender.

Paris' efforts are impressive – as are the volumes of animal excreta left on its streets – but to really learn about toilets, you have to head east. Clara Greed, a professor of planning and author of *Inclusive Urban Design: Public Toilets*, tells me she envies the Asians. They are properly thoughtful about toilets. (At WTO events, it is common to find yourself surrounded by Singaporeans, Malaysians and Hong Kong Chinese, who impress you with the size of their delegations and the seriousness of their intent.) Asians have come to understand that public toilets have a value that is both moral and monetary. It is ironic that one of the best examples of a well-funded and well-executed public toilet programme, one that would make British Toilet Association members sigh with envy, has happened in a country better known for toilet standards that are more execrable than exemplary.

It is raining in Beijing. Straight, hard, determined rain. I welcome the change. I'd been here five days before I saw the sun, though every day had been sunny behind the smog. Beijing is a city where the weather doesn't seem to matter, like it has lost hope. Even this rain may be suspicious. It's August 2006, two years away from an Olympics that the Chinese authorities are determined will be a showcase to the world. An English ex-pat tells me that everyone suspects the authorities of seeding the skies to make rain, to wash away pollution. Officials denied it, but the rain had still arrived every night in July, at 10 p.m. exactly. Beijing's rulers want to clean up the city for the Games and if rain-showers work, they'll try it. They also want to clean up the city's restrooms for the expected three million visitors. The clean-up plan – launched in 2005 – had epic ambitions. Five thousand public toilets would be constructed or renovated. No-one would be more than five minutes' walking distance from a decent toilet. Any traveller to China, used to facilities that are few and filthy, might find these goals as unattainable as trying to tame the weather.

Top-down efforts to 'civilize' China are not new. The New Life Movement of Chiang Kai-Shek, the Kuonmintang leader defeated by

Mao's Communist Party, laid out ninety-six specific rules to improve Chinese virtues. Better behaved Chinese would herald a stronger, more competitive modern nation. The rules included no smoking, sneezing, spitting or urinating in public. Eighty years on, China's authorities are trying an equally daunting social programme, only this time the targets are China's people and its toilets. Civilizing campaigns launched over the past few years have been aimed at eradicating spitting, bad manners, rude taxi drivers, bad English, and flies. Sadly, this civilizing campaign will get rid of such Chinglish signs as 'Deformed Man', to indicate a handicapped toilet, and 'Show mercy to the slender grass' signs on Beijing's parks. But it has introduced Chinese readers to the inimitable Guo Zhangqi, a farmer who travels every day to Beijing to stand in a public park offering $2.50 for five dead flies, and funds his goal of 'a fly-free Olympics' out of his own pocket.

The construction and renovation of 747 of the 5000 public toilets was handed to the care of Mr Pang of the Tourism Development Authority. His office is in a building that has a TV in the lobby showing a video of some of China's lush scenery – green rivers, clear rushing waters, not a smokestack in sight. Mr Pang is a middle-aged man in short-sleeves who hands out goodie bags of Tourism Development Authority bags and literature, and shows us into a meeting room with plenty of windows through which to watch the teeming rain. Mr Pang isn't responsible for toilets any more, which may be why he agreed to meet us. (When I try half-heartedly to meet with someone in the Beijing sanitation department, I am instructed to contact the Propaganda Ministry. I don't.) The 747 toilets have been constructed and renovated, and the programme is deemed to be completed.

Still, he's happy to talk about them. They are something to be proud of, and they are essential to his job. 'Tourism is a window industry. We want foreigners to have a pleasant experience. Toilets are so important: without them, tourists don't come back.' The Tourism Development Authority decided to provide squatting designs and sitting ones, to be accommodating to all preferences. There are 'third space' toilets which families can use together. The construction was accompanied by public campaigns exhorting Chinese people to close the cubicle door, or to use less water, to behave more like civilized people. For a proud nation, it seems odd that the Chinese are choosing Western definitions of civilization to form their own. 'In reality,' Norbert Elias wrote, 'our terms "civilized" and "uncivilized" do not constitute an antithesis of the kind that exists between "good" and "bad".' Concepts of what civilization means are more slippery than that. Chinese newspapers may be urging their citizens to be better behaved – and they mean more like Westerners – but it cuts both ways. Westerners could learn some manners too, though Mr Pang is too polite to say so. Western habits of putting wads of toilet paper – and plenty of other things – down the toilet coincide unhappily with China's smaller sewers, which clog easily. I ask Mr Pang how he's going to educate visitors to put toilet paper in a basket like the Chinese do, a practice that according to Western concepts of civility is dirty and unhygienic. 'I don't know,' he says. 'We can't follow them into the cubicle.' The Chinese can harness rain-showers and turn foul toilets into gleaming ones worthy of Olympic ideals, both considerable achievements. But they can't dictate toilet behaviour, any more than anyone can, any more than sparrows can be tamed or flies bought into extinction, $2.50 for five.

There is no secret to how the Chinese overhauled their public restrooms. There was money, it was made available and toilets were built. A command economy and an authoritarian political system helped. In the world's most famous democracy, things are handled very differently. In the US, a country notable for its inability to provide acceptable levels of away-from-home toilets, the public's reaction to a lack of a fundamental public service is generally, like, whatever. The American Restroom Association wasn't formed until 2004, and has yet to make much impact. The Privy Council, a New York City-based toilet pressure group, hasn't updated its website for years. Apart from that, there is nothing.

Perhaps such meekness is due to urban-dwelling stoicism. During a blackout that was an indictment of an imperfect infrastructure, New Yorkers threw parties. Perhaps it's due to language: a country that chooses 'restroom' and 'bathroom' to signify places that dispose of human excreta may not want to look beyond the language barrier. Perhaps it's because humans are adaptable. When there isn't a toilet, people manage, or they stay home. In 1966, Alexander Kira could crack the old joke about the difference between a camel and a lady: a camel can go all day without drinking, and a lady can drink all day without going. Self-restraint is the foundation of propriety, but this much restraint has serious health consequences. One organization which lobbies for older people's rights refers to the 'bladder leash', which confines hundreds of thousands of elderly people to their homes because they are scared of not being able to find a toilet when they leave the house. (They call it a bladder leash because 'bladder and colon leash' wouldn't get any publicity.) Bladders can be shy as well as leashed. 'Shy bladder syndrome', as parusesis is usually called, affects one million Americans, according to the International Parusesis Society, and renders them incapable of urinating in public places. Incontinence, meanwhile, affects 25 million Americans, according to the National Association for Continence, but you'd never guess from their minimal public profile. Compare this to Australians, not known for their sensitive nature, who in 2001 launched a national toilet map to help their incontinent and continent citizens find the country's 13,000 public restrooms. Even the toilet-closing English of Westminster Council have now launched Satlav, a service that can send text messages to subscribers with the location of the nearest convenience.

Americans can get excited about bathrooms, but – unsurprisingly in a litigious country – only in the law courts. The biggest bathroom battle isn't about the absence of a common decency, but about inequality. The 'potty parity' movement is led by a lawyer called John Banzhaf, who made his reputation by founding the anti-tobacco lobby group Action against Smoking and Health (ASH). He objected to smoking because it damages other people. It is unfair. He came to potty parity for the same reasons. Banzhaf seeks to better an age-old truth of public restrooms: women always have to wait. There are always lines in restrooms because it takes longer for women to pee. Careful research has established that women take ninety seconds to urinate, while men take forty-five. Men who complain that women should hurry the hell up do not take into consideration the fact that women have to undress and sit. They have cumbersome clothes. They have shopping bags sometimes, and small children. Providing an equal number of restrooms for men and women, Banzhaf argues, doesn't help, especially when square footage is taken into consideration: more urinals can be fitted into the same floor space.

Banzhaf first took the case of Jean Ledwith King, who filed a complaint against the University of Michigan, which was going to renovate an auditorium and install twenty-two men's stalls and thirty for women. More recently, he filed a complaint with the Architect of the Capitol, claiming that 'the failure of the House to provide [...] equivalent access for women constitutes illegal sex discrimination and violates the constitutional right of Equal Protection'. Also, apparently it makes them miss the vote. Banzhaf's press release highlighted earlier research which found that male members of Congress have access to a toilet a few feet off the House floor. This includes 'six stalls, four urinals, gilt mirrors, a shoeshine, ceiling fan, drinking fountain and television'. The seventy female members of Congress, meanwhile, should they need the toilet, would have to traverse 'a hall where tourists gather, or [enter] the minority leader's office, navigating a corridor that winds past secretarial desks and punching in a keypad code to ensure restricted access.'

There are other routes to equality that don't involve suing. Unisex restrooms are often touted as a solution to potty disparity. The unisex restroom that starred in the American TV show *Ally McBeal* became as famous as its human stars. But unisex toilets are not popular. They make it harder to practise civil inattention, to pretend. Our feelings towards public toilets are already more negative than towards home ones, as Kira wrote. We don't want to see faecal matter in the toilet bowl, which is too much information about another person, and we don't want a warm seat, which is a sign that a stranger has left his/her body heat, and from a bare body, too. That, says Kira, 'is more sharing than many people feel comfortable with'. Public toilets are only tolerable when users can pretend that they have 'mineness'. 'In a relatively spotless public bathroom, with no-one "passing wind" or whatever, it is perfectly possible for us to pretend we are in a private situation – in a bathroom, or booth, that is "mine".'

A unisex toilet not only has the stranger problem of regular public toilets, but contravenes the social codes of gender segregation that have prevailed for ever, and that operate with more or less severity, in most countries in the world. Even in rural China, where I turned up unannounced in a tiny village in the middle of mountains, by one of those rushing rivers from the Tourism Development Authority video, and questioned a thirteen-year-old girl called Chen Xie about her family latrine. She was charming and showed no alarm about strangers arriving to ask her about her toilet, until I asked her what happened when their latrine – which had no door, this being China – was occupied and someone else came to use it. 'If it's a woman,' she said, 'they can come in.' And if it's your father? She looked appalled. 'No! That would be awful.' Then she led us back to the house and her grandfather offered a plate of watermelon, and though I'd just been told that the family fields were fertilized with the contents of the family latrine, I took the fruit. Watermelons have thick skins.

There is some innovation and invention in the public toilet field. Germans, for example, think urinals are flawed. Aiming a stream of urine at a toilet bowl sends a fine spray around the room (as does every toilet flushed without the lid closed). Spray becomes vapour, which leaves a chemical deposit on anything surrounding the urinal. It can also change the colour of wallpaper. The Japanese toilet firm National dealt with this by putting a dot of light in the bowl to serve as a target, a concept developed after staff at Tokyo's main airport noticed that putting stickers in the bowl improved men's aim and kept their floors cleaner.

The Germans want men to sit instead. I discovered this curious cultural fact when a friend had a relationship with a German woman, who found his habit of standing to urinate as odd as he found her insistence that he sit down. She did not thankfully resort to the cheap technological answer to her problem, a ten-dollar German-made alarm that is attached to a toilet seat. When the seat is lifted, the transgressor is admonished that 'stand-peeing is not allowed'. But the relationship didn't last and my friend now stands unimpeded, and probably cherishes his copy of Klaus Schwerma's interesting book *Standing Urinators: the last bastion of masculinity?*

The other way goes the other way. Men should sit down, and women should stand up. This is not an outlandish concept: women have stood to urinate at various times and places throughout history. Herodotus reported that Egyptian women 'stand erect to make water; the men stoop'. The grand ladies of the French court would pee and defecate at will with no shame, but they had clothing that was better suited to standing. The modern woman, with her trousers and underwear and shopping bags, is unlikely to embrace the female urinal, and hasn't yet. There have been pockets of success – Swedish pharmacies do a creditable trade in cardboard funnels that enable women to stand-pee, and cardboard funnels and She-Pee female urinals were well-used at the 2007 Glastonbury music festival. But TOTO's female urinal, launched in 1964, failed miserably. I saw many of these curious objects in Japanese department store toilets, and no-one was going near them and I wouldn't either.

Female urinals are often offered as a solution to restroom congestion, and they are usually offered as a solution by men who won't have to use them. For this reason, they really enrage Clara Greed. When a pleasant enough Hong Kong doctor suggested they are a good idea, during the WTO summit in Moscow, Greed was moved to leap from her seat and say, 'This is war!' She hates female urinals because the concept is short-sighted. Women would not take as long if their public toilets were better designed. There would be fewer queues if there were more public toilets. All sorts of problems could be solved if only people in charge got it into their head that providing their citizens with public toilets was not only the height of civility, but good economic management. In rural Australia, development thinker Peter Kenyon, who runs the Bank of I.D.E.A.S. institute, promotes toilets as the means to prosperity for the thousands of two-horse, public-toiletless settlements that are trying to survive. Because he is Australian, he calls this the 'shit-led revival'. By his calculations, settlements located on good roads can get an extra twenty carloads of people to stop every day if they install a public toilet, which would mean as much revenue as a smallish rural factory would make. Restaurants and service stations know that decent toilets bring in trade, and outlandish toilets – ones that play language tapes, or have special glass that means you can see out, but that people can't see in – bring the media, which bring clients. But such effort is made in the private sector and nowhere else.

Public bathrooms, and their absence, throw up big questions. For me, the questions started with the doors. In China, I asked a lot of people about why their toilets had none. I thought I was asking a question about habit and design, but without exception, I was given answers that were about civility. A Communist Party official I met in Chengdu was offended by the question. 'Those are the old style toilets. We are civilized now.' Wang Ming Ying hadn't thought about it before, but now she did, she kept her options open as to the reason, musing that 'maybe it's an indication of a lack of civility. Or because when we go to the toilet we are all the same?' An ex-pat in Beijing said, 'All my colleagues leave the door open. It's because they're not bothered.' For the Chinese, civilization is not about privacy or enclosures. They have public bathrooms with no doors; but those civilized Westerners have hardly any public restrooms to put the doors in.

Those Chinese who are being publicly educated to adopt Western standards of civility might reasonably point to a certain rather bizarre toilet-related scandal of 2007, when Senator Larry Craig of Idaho pleaded guilty to disorderly conduct with an undercover police officer in an airport toilet stall. The case generated plenty of media glee about homosexual pick-up codes such as foot-tapping. But I waited in vain for someone to explain the weirdest thing about the whole story: that Craig had been able to peer in at the cop on the toilet because nearly all American public toilet partitions have gaps in the doors large enough to see into. New York University sociologist Harvey Molotch was surprised when I asked him about it. He'd never thought about it but now that I mentioned it, 'I suppose it's about control.' The need to control criminal activity overrides prudish values that were made obvious when the owners of the offending Craig airport toilet announced they would install 'chastity partitions', with no irony. By conventional standards of civility, American public toilets are caught short.

Does that matter? Even in public-toilet-deprived cities like London and New York City, there are toilets available, even if they are in cafés or pubs, and even if they cost a muffin or a cake, even if they don't have chastity partitions. But those who make a living of caring about public bathroom provision think toilets are a test of what living in a city means, of what civilization is. Concepts of civility and propriety are complex and changing – there are now CCTV cameras in some English pub toilets – and so is the public toilet. It's a work in progress along with the civilization it is supposed to represent, a truth immortalized in the excellent one-liner by Mohandas K. Gandhi when he was asked what he thought of Western civilization. He replied, 'I think it would be a good idea.'

Blue Plains wastewater treatment plant, Washington, DC – *Author*

7. THE BATTLE OF BIOSOLIDS

Bad Smell, Big Tomatoes

It is my first and last day of sewage school. The school premises are nothing much to look at, consisting of a Portakabin in the car park of Barston sewage treatment works near Birmingham. I've joined a class of local schoolchildren, in one of five classrooms run by Severn Trent, one of ten utilities that supply clean drinking water and remove dirty water for the people of England and Wales. The education programme is funded by the utility in an attempt to remind the public of the vital work it does. It is considered a good investment.

After a brief trot through the water cycle and some green lessons – washing a car with a hosepipe uses nine litres of water a minute, children, so use buckets – we put on our wellies for the tour. The facility dates from 1911 and hasn't changed much since, but it has all the basics of modern wastewater treatment. First, the influent, brown water rushing in from the sewers, that can be seen through a viewing hole in the ground. There are the screening grills, which catch all the large foreign objects, then the compactor, which crunches them up.

It looks like it's not working, but only because its pace is glacial. Slowly, slowly, it spits out rags and pen caps and thousands of yellow sweet-corn kernels that humans can't digest. They dot the muck like gems, and are prized by birds.

Next come the first tanks. Wastewater treatment is much tinkered with – a thousand treatment works will have 999 different processes, a worker tells me – but the basics are unchangeable. Solids are removed from sewage first by filtering and letting them sink. This is primary treatment. Secondary treatment involves feeding oxygen to micro-organisms that break down any organic content still in the wastewater. Tertiary treatment cleans the water further with sand filtration or UV light disinfection.

At the settlement tanks, we learn that the scum on the top of the water is grease that won't sink, and that the heaviest solids are already settling to the bottom. A condom – an 'adult balloon', in the teacher's whispered words – floats near the surface. Secondary treatment here is done with old-fashioned bacteria beds, huge circular tanks filled with bits of coke, limestone or special plastic parts that are covered with bacteria and sprayed with sewage water from endlessly rotating mechanical arms. The bacti-beds have fallen out of favour because they tend to be smelly. Aeration is a newer, better way of getting oxygen to the bacteria so they eat faster. Elsewhere, this is done in long lines of tanks known as aeration lanes, where the brown waters foam from the bubbling oxygen. (A chocolate mousse effect is desirable.) Secondary treatment is all Barston does, and the bacteria-cleaned effluent goes into a nearby stream. I lean over obediently to look at its colour. The teacher enthuses. It's clear! Not brown! And then it's time to make sewage soup.

It is centuries since sewage consisted of pure human faecal material. Into sewage, anything goes. The French call this *tout à l'égout*, or everything down the drain (to be contrasted with their earlier, smellier habit of discharging *tout à la rue*). An enterprising sewage treatment manager in Utah arrestingly expressed this by issuing bottles of cleaned sewage effluent, whose labels listed the following ingredients: 'Water, faecal matter, toilet paper, hair, lint, rancid grease, stomach acid and trace amounts of Pepto Bismol, chocolate, urine, body oils, dead skin, industrial chemicals, ammonia, soil, laundry soap, bath soap, shaving cream, sweat, saliva, salt, sugar. No artificial colours or preservatives. Some variations in taste and/or colour may occur due to holidays, predominant cuisine preference, infiltration/ inflow, or sewer cross-connections.'

The manager told reporters that he was trying to raise people's consciousness about what they put down the drain. In Severn Trent's classroom, the same message is conveyed by the teacher, who has taken sewage class more times than he can remember. Each time he watches the same reactions: Disgust. Amusement. And if he's lucky, enlightenment.

The ingredients of sewage soup are a tankful of water and whatever else the class might have put down the sink, toilet, gutter or drain that day. The teacher asks for suggestions, then adds the contaminant to the water. Shampoo, soap, toothpaste, washing powder, rice, salt. A 'number one' is lime cordial. A 'number two' is Weetabix soaked in water. That's the easy part.

The rest of our lesson involves filtering the filth out of the water, in an attempt to impart the difficulty – and dubious sanity – of modern industrial society's paradigm of waterborne waste treatment, whereby you take clean drinking water, throw filth into it, then spend millions to clean it again. My team manages to get a passable liquid. The teacher is pleased. But nobody has considered the stuff that's been filtered. Nobody mentions the sludge.

When sewage is cleaned and treated, the dirt that is collected and removed is called sludge, except in the US, where it's called biosolids by some people and poison by others. For the last twenty years, the blandly named product has been the centre of an increasingly loud and bitter battle. The great Victorian gardener Sir Joseph Paxton once said of sewage that it was 'a great rough sort of business. You cannot put it into nice forms and ways.' In the US, the biosolids debate – and debate is a polite way of putting it – has reached the highest corridors of power. It has made people rich. It has come to involve alleged deaths and illness, lawsuits and bitterness. In short, sludge disposal has become a very rough sort of business.

I contact the official face of the biosolids industry in the US. The National Biosolids Partnership, an alliance between EPA and the wastewater industry, directs me elsewhere. They want me to see the best of the best, and the obvious choice is a small facility half an hour from the US Capitol building.

The town of Alexandria has cobbled streets and expensive town-houses, colonial aesthetics and money. We are getting lost, and it's my fault. The taxi driver has never been to the Alexandria Sanitation Authority treatment plant before. He thinks it should be near the Old Town, but I say that's impossible. Wastewater treatment plants are nearly always on the outskirts of town. Anyway, I'd be able to locate the place by the smell, with which I was by now familiar: mustiness, on the edge of unpleasant but not quite. I was wrong on all counts.

A few blocks west of the expensive town-houses, we find the smartest, sweetest-smelling sewage treatment plant I've ever seen. I'm expecting to be met by a middle-aged engineer, as wastewater treatment senior staff generally are, but Paul Carbary is a young, handsome man with a goatee, a subtle tattoo and a cheerful manner. Though he looks youthful enough to be in college, Carbary directs ASA's biosolids programme, which has the optimistic name of Green Fields. He is one of the public faces of biosolids, as is ASA, and in these times, the industry needs all the good PR it can get.

Carbary is joined by the team leader of de-watering (a process that makes soggy sludge less soggy), a terse, big man called Joel Gregory, who looks like he'd rather be dealing with sludge than with me and only reveals a darkly dry sense of humour after an hour or so of silence. Carbary, though, is all energy and bounce. He has an earnest eagerness which I will let flow over me, because it's early and they have provided breakfast.

They launch into a lecture on the process of biosolids production, officially known as 'the solids treatment train'. The lecture is designed for ASA trainees, but Carbary translates the technical terminology so I get the essence. The solids removed from liquid sewage are first thickened then treated. The goal is to take a liquid product that can be dangerous and make something dry and safe enough to apply to land as fertilizer. At ASA, solids are thickened using centrifuges that remove water content – 'kind of like a washing machine, throwing the liquid out to the sides' – then treated in pasteurising heat facilities that 'jack the temperature up to 70 degrees Celsius, hold it, and kill all the disease-causing organisms in the sludge'. What, salmonella and E. coli? Helminths and ascaris ova?

Carbary is confident. 'Yes. Gone.'

Sludge is classified according to the Environmental Protection Agency's Code of Federal Regulations, Title 40, Part 503 (usually shortened to 'the part 503 rules'). Class A has to be treated until no pathogens can be detected; in Class B, pathogens are reduced but still present. Class A Exceptional Quality, which ASA produces, must have fewer heavy metals than Class A. This is more expensive to produce, but can be applied more liberally: EQ solids can be spread in parks, fields and school yards with no requirement to inform neighbours or local officials. Class B can be applied to agricultural land, reclaimed sites and forests, all judged to be 'non-public contact' sites, with some restrictions. (Sites must be a certain distance from human habitation, and only certain crops can be harvested.) It can also be sprayed on trees.

Carbary and Gregory are proud of their facilities. They should be, because $45 million was spent upgrading the plant to produce EQ. Those millions have funded a process that in Carbary's view is 'pretty damn cool'. He wants to convey how cool it is by using an analogy I can understand. 'It's like turning milkshake into cake.'

Gregory objects. 'Not milkshake. It's more like coffee.'

'But it's 2.5 per cent solids. It's milkshake!'

'Coffee with grounds in it then. But definitely coffee.'

An inspection will settle it. They give me a hard hat bearing a US flag and a 'United We Stand' logo. They overlook my open-toed shoes, which are forbidden, and united we head out onto the campus. It's called a campus because it looks like one. The buildings are red-brick and compact. It has a college quadrangle neatness. I've never seen any sewage works like it, because they are usually huge and spread out and dotted with unsightly settlement tanks or aeration lanes. Sewage treatment is a messy, malodorous business, but not here. ASA built this plant to blend in aesthetically with the old-worldness of Alexandria, and with the city's new-world sensitivities to smell.

Odour is the biggest headache for a treatment plant operator. It is the cause of most public complaints. The huge Mogden works in west London even generated a name for its odour – the Mogden Pong – and a combative residents' group that runs a site to collect residents' complaints. One email reads, simply, 'Stench'. Water utilities spend millions on odour treatment, and are regularly taken to court by local government for breaking nuisance laws. Saying that 'sewage is bound to smell', as one utility manager told me, is not sympathetically received by people who must cancel their garden parties when the wind blows.

Another wastewater manager told me it was people's own fault. 'Who'd buy a house right next door to a wastewater plant? Were you drunk?' But houses arrive when cities expand. It's not the fault of the sewage works when the city grows to meet them. Sewage treatment professionals think odour complaints arise from ignorance. Joel tells a tale to illustrate this. 'My brother's a cop. His wife does my taxes. I make more money than my brother, so my sister-in-law says, "How in the world is that right, that you make more money than a policeman?" She was sitting down drinking a glass of water. I picked up some dirt and dumped it in her water and said, drink it now. If I don't work, eventually that's what you're going to drink. If we don't pre-treat that water, water treatment plants can't treat it for you.' She never queried his salary again.

ASA spent millions on an odour treatment removal system. But the 'jewel in the crown' is a 115-feet building that houses solids treatment. I'd been told by another wastewater manager that I'd be able to eat off the floor here, and he wasn't lying. Other sewage treatment plants invariably had parts that made me retch but this place is spotless and odourless. We traipse through the several floors following the sludge through its transformation. The trucks pick up the EQ result on the ground floor from 5 a.m. onwards, early enough to avoid the traffic and not disturb the neighbours, to take the biosolids to fields in Virginia. ASA pays a company called Synagro to take its product away. Synagro – which describes itself as 'a residuals management company' – is paid by a lot of states to take sludge away, and then to apply it to land, turn it into pellets, or landfill it. It's good business. Ten years ago, according to a *Houston Press* investigation, the company operated in three states and had an income of $20 million. Now, it's active in thirty-three and has revenues of $320 million.

Joel takes some sludge from a chute and tells me to do the same. It's black, earthy. It looks and feels like a crumbled brownie, rich and fertile. Farmers love this stuff, says Joel. They honestly can't get enough. It's not as good as artificial fertilizers, which have more nitrogen and phosphorus, but for a free product, it's unbeatable. I ask him if he'd use it on his garden, and he knows that I'm actually asking if it's safe. 'Sure I would. And I'd have my kids roll around in it. No problem.' If they ever bag it and sell it in garden centres, he'd be a satisfied customer. Why not? It's a marvel, an expensive and remarkable transformation. Cake from milkshake. It's nearly as good a transfiguration as turning 'sludge' into 'biosolids'.

By 1992, American sewage treatment plants had a problem. For the ninety or so years since most cities had installed sewers, the way human waste had been dealt with had not evolved. Remove sludge from the liquid sewage, then treat the water and dump the sludge at sea, in rivers or in landfill. Effluent from industrial sources was dumped with little regulation. By 1969, *Time* could write that the nation's rivers were 'convenient, free sewers'. It described Ohio's Cuyahoga River, which caught fire twice that year, as 'chocolate-brown, oily, bubbling with sub-surface gases'. Consequently, the Clean Water Act of 1972 provided big money for municipalities to improve their sewage treatment.

The construction and renovation frenzy that ensued was the largest public works project in the country to date. By its completion, the US had 16,000 sewage treatment plants, and an improved sewage treatment process. But cleaning sewage more efficiently meant removing more dirt. In other words, the Clean Water Act increased the amount of sludge being produced, which was mostly dumped at sea. Farmers like sludge because it has nutrients, but the same nitrogen and phosphorus can feed and breed algae that suck out water's dissolved oxygen content, leaving it lifeless. Sewage can suffocate the sea. After too many toxic shellfish beds and algal blooms, Congress passed the 1989 Ocean Dumping Ban Act. This gave the industry four years to come up with an alternative to ocean dumping. Americans produce 7 million dry tons of sludge a year. It had to go somewhere.

At this point, someone must have remembered the 'sewage

doctors' of nineteenth-century Europe. Men like Justus von Liebig and Alderman Mechi thought Joseph Bazalgette's new sanitary sewer system was a criminal waste of a potential fertilizer because it discharged the contents of the sewers into the Thames. Liebig calculated the worth of this lost wealth at £4 million. Karl Marx remarked that London 'could find no better use for the excretion of four and one-half million human beings than to contaminate the Thames with it at heavy expense'. Alderman Mechi, luxuriating in the modern convenience of household-supplied gas, imagined a future 'when each farmer will turn on the tap and supply himself with town sewage through his meter according to his requirements'.

Sewage taps never caught on, but until wastewater treatment became pervasive in the early twentieth century, sewage farms, which irrigated their fields with raw sewage, thrived all over. Gennevilliers, outside Paris, produced vegetables highly sought after by Parisian restaurateurs. The sewage farm at Pasadena, California, grew excellent walnuts. The arrival of cheap artificial fertilizers made sewage farming uneconomical, but the principle was sound. Properly treated, sewage could have a place in the nutrient cycle. Food feeds humans whose waste feeds food.

But by end of the century sludge contained far more than pure human excrement, and hardly any of it good. Anything that gets into the sewers can end up in sludge. US industry is estimated to use 100,000 chemicals, with 1000 new chemicals being added every year. These chemicals can include PCBs and phthalates, dioxins and other carcinogens. Sludge may contain pathogens from all sorts of sources. Hospital and funeral waste can include SARS, TB, hepatitis. Sick people excrete sickness, and it all ends up in the sewers. The *Harper-Collins Dictionary of Environmental Science* defines sludge as 'a viscous, semisolid mixture of bacteria and virus-laden organic matter, toxic metals, synthetic organic chemicals and settled solids removed from domestic industrial wastewater at a sewage treatment plant'. The Clean Water Act keeps it simple and calls it a pollutant.

I have to use words like 'if' and 'may', because no-one actually knows what's in sludge. Technically, industries are supposed to pre-treat hazardous chemicals and waste, but oversight is minimal. And anyway, no-one regulates how thousands of chemicals might react with one another, or with the pathogens floating alongside them. The most optimistic view of sludge is that it is a soup of unknowns. Others think it's toxic and can't be anything else: cleaning water is done by removing contaminants and concentrating them in sludge. The better the wastewater treatment process, the worse the sludge.

So its transformation into fertilizer was going to be a hard sell, but there was little alternative. There are five options for sludge disposal: landfill, incineration, gasification, disposal at sea, and land application. The first three are costly and ocean dumping is illegal. Land application is legal and cheap. It was not a difficult choice. Also, there was precedent. In Wisconsin, the Milwaukee Metropolitan Sewerage District (MMSD) has been selling its sludge as fertilizer – brand-named Milorganite – since 1925, with discreet labelling. Only someone who knew what MMSD stood for would realize Milorganite is derived from a human body.

The sludge and wastewater industry looked at Milorganite, and saw the light. No-one would want to live near farms where sewage sludge was applied. But people might want to live near fields that were covered in a fertilizer called something else. The transformation of sludge into 'biosolids' was brilliantly documented in *Toxic Sludge is Good for You* by John Stauber and Sheldon Rampton. The book was about 'the lies, damn lies' of the PR industry in general, but the manoeuvres of the Water Environment Federation (WEF), the US sewage industry association, were impressive enough to provide the authors with their title. The EPA, they write, was conscious even in 1981 of the need to persuade the public to accept sludge farming. A

Name Change Task Force was formed, and suggestions solicited through a WEF newsletter. The 250 suggestions received included 'bioslurp', 'black gold', 'the end product', 'hudoo', 'powergro', and – my favourite – R.O.S.E, standing for 'Recycling Of Solids Environmentally'. Biosolids won, probably because it was the blandest. Maureen Reilly, a prominent sludge opponent and the producer of the prolific *SludgeWatch* newsletter, calls this 'linguistic detoxification'.

Chris Peot thinks the name change was common sense and nothing more. He uses Jell-O as an example. 'It's called Jell-O for a reason. It's not called gelatinous red goo because no-one would buy it. Of course you're going to try and have a name for your product that is palatable. This is a product that helps the farmers, that is valuable. We'd be insane to call it sludge.'

Peot's was the other name suggested by the National Biosolids Partnership. He runs the biosolids programme at Washington, DC's Blue Plains plant. He's on a conference call when I arrive at his office, so I pass the time reading the situations vacant advertisements in the lobby. A young man is filling in an application form, and I wonder if it's the Position Vacancy Announcement for a Wastewater Treatment Plant Operator, a job which will require him to 'troubleshoot operational problems including process microbiology' and 'perform laboratory analysis'. Wastewater personnel keep effluent clean. Because effluent ends up in drinking water sources, they are also the guardians of the nation's safe drinking water. To qualify for this weighty responsibility, the Operator is required to have a high school diploma and lift fifty pounds.

Blue Plains is so big, it has road-signs. (I like 'Solids Road' best.) Despite its vastness, it is totally automated, and directed from a command centre which looks like mission control, with three enormous plasma screens and several banks of monitors. The place is impressive, and so is Peot, when we finally coincide. He is handsome, like Carbary, tall and lean and dressed in normal clothes – hiking boots, comfortable trousers – not a suit. I'm beginning to wonder if all biosolids managers are designed by Gap.

He is also languid and likeable, but the calmness shouldn't deceive. He is in charge of an operation worth $20 million a year. He has a $250,000 annual research budget. I start with the most obvious question. Why do people hate biosolids so much?

He thinks it's all about smell. 'We're all pre-programmed to be afraid of our own waste.' Cave people who didn't avoid the potentially pathogenic matter that was their own excrement probably got sick and died out. 'Maybe they slipped and fell in it, then they went back into the cave and made a sandwich and they culled themselves from the gene pool. Everyone who was sensitive to the sulphur odours, which is the trigger, avoided it. So you have a population that's extremely sensitive to sulphur odours.' Sulphur has been shown to produce anxiety, he says. 'That's exactly what we're producing from biosolids. So we have a hypersensitive population which is here only because they knew how to avoid the faecal-oral cycle. They're getting this smell of sulphur and then when they find out it's human it's even worse.' Women are the most vocal protesters because they are the protective ones. It's primal.

This is a good argument because it is hard to argue with it. Odour can be measured with olfactometry machines – which can bag an odour and then have it smelled by testers – but it's hard to be definitive about a largely subjective reaction. In 2004, the grand jury in Orange County, California, in a report entitled 'Does anyone want Orange County Sanitation District's 230,000 tons of biosolids?' noted that biosolids proponents described their product's smell as 'musty' or 'earthy' whereas opponents preferred words like 'noxious, horrible, putrid, nauseating, eye-watering and sickening'. The jury quoted a list published by the National Academy of Sciences of odorous compounds found in sludge. These included hydrogen sulphide (which smells like rotten eggs); dimethyl sulphide and carbon sulphide (decayed vegetables); thiocresol (skunk-like odour); methylamine, dimethylamine and trimethylamine (fishy); pyridine ('disagreeable and irritating'). In all fairness, they concluded, 'acetaldehyde is reported to smell like apples'. The report writers could 'only imagine what odour might emanate from a concoction of these compounds. "Musty" or "earthy" doesn't come to mind.'

Peot has regular olfactometry panels to test the palatability of his biosolids. He says that if there's something wrong with them, he wants to know about it. He speaks the language of co-operation. He's on a panel with Ellen Harrison, director of the Waste Management Institute at Cornell University and a known sceptic of biosolids. He is funding research by Dr Rob Hale, a microbiologist at the University of Virginia who found that organic chemicals found in flame retardants persist in sludge despite treatment and end up in the food chain.

He regularly attends public meetings where the safety of biosolids is debated, usually ferociously. He thinks it's important to show the friendly face of biosolids. 'If it's Synagro [in attendance] or some other big public behemoth, people aren't going to believe it.' He is more believable, he says, because he has less at stake. 'I'm a public servant. I'm doing this because I think it's helping to save the planet in my little corner of the world.'

In another little corner of the world, Peot's routine would persuade no-one. In this corner, a brown field covered with Class B biosolids sits a hundred feet or so from Nancy Holt's white bungalow. The field is a six-hour drive from Washington, through Virginia and North Carolina, through an invisible border where all the radio stations turn into country music ones, off Route 85 to a quiet hamlet near the town of Mebane.

I arrive sweaty and dazed by jingles and guitar jangling to be greeted by a warm woman who gives me a hug and a wet rag for my forehead. Her accent – treacly Southern – soothes too. Nancy lives here with her husband Bruce, who wears his white hair long and Willie Nelson–like, and blue topaz on his fingers. Family photos of pretty granddaughters and handsome sons are placed around the tidy household. On a car outside, there are anti-war stickers. This is picture-postcard rural America. Except some things aren't visible in the picture, such as the effects of the farmer's fertilizer over the road, and what Nancy Holt thinks it has done to her and her neighbours, which is nothing good and possibly lethal.

Nancy has lived in this hamlet all her life, and her family has farmed land in these parts for 250 years. Nancy wasn't interested in farming the land, though she still lives on it. Instead she became a nurse, then moved into medical equipment sales. She serves iced tea, and then some more, then sits me down at the kitchen table and prepares the weapons of the grassroots protester: piles of files, dossiers, reports, and a scientific vocabulary that she has accumulated along with frustration and disbelief. She begins by saying that the year before, sludge was applied for thirty-three days straight to the field in question. 'Based on the number of 6000-gallon tankers that came to apply it, we came up with the best guess that 9.75 million gallons [were] spread on 160 acres. They were doing it 12 hours a day and a truck would arrive every 10 minutes.' That was when she went blind.

She wasn't a well woman to begin with. When I'd called to make the appointment, she'd apologized for misunderstanding something by saying, 'I have holes in my head.' I took it as a joke, but she does have holes in her head, after surgery for an ailment that she doesn't describe but which left her with metal clamps in her brain. One time when the sludge was applied – it's been arriving twice a year, spring and summer, for thirteen years – the arteries in her brain swelled, pressed on her optic nerve and temporarily took away her sight. The diagnosis was giant cell arteritis, but no cause was proven. Nancy is sure the cause was sludge, and she now spends much of her life trying to prove it.

The stuff on the field over the road is not the fancy Class A EQ-pasteurized and digested stuff coming out of Alexandria's expensive treatment processes. Class A may be biosolids' public face, but most of the 3 million dry tons of sludge applied to American farmland is Class B, whose face is uglier. Under the Part 503 rules, Class B has to be stabilized somehow (adding lime is popular) and that's it. Though an EPA study in 1989 found twenty-five groups of pathogens in sludge – E. coli, salmonella, worms and fungus – it decided to regulate only nine heavy metals. The power of soil, sun and general degradation, it was thought, would take care of the rest. Soil is a complex and delicate eco-system, and there is much about it that is still unknown. But by other countries' standards, the EPA's were lax.

When Cornell University scientists Ellen Harrison and Murray McBride appraised the Part 503 rules in a 1999 paper, they compared the US regulations to a more precautionary policy prevalent in many European countries. There, farmers shouldn't add to the levels of heavy metals already in soil. The US, by contrast, uses a risk-based approach. It calculates what soil – and people who live near to and work on the soil – can handle. In practice, according to Harrison, farmers can apply sludge until their crop yield is reduced by fifty per cent. In other words, they can proceed until harm is shown to have been done. Harrison attributes this philosophical difference to land and time. The US is bigger. 'We're used to being able to screw it up and just find a new piece of land. Europe is more densely developed. There's no outback. They have to take a precautionary view.' And Europeans know that lead used by Romans two thousand years ago is still in their hills. They can't afford to calculate that soil can handle risk; they prefer not to risk that it can't.

The EPA has stated that its Part 503 rules make the application of biosolids 'an appropriate choice for communities' as long as certain limitations are followed. These regulations include buffer zones and other 'management practices' that are supposed to determine how, when and what quantity sludge can be applied to soils. In North Carolina, land used to grow crops that touch the top of the soil – cucumbers, melons, tomatoes – must not be planted for fourteen months after sludge application. For crops that grow beneath the soil it's thirty-eight months. At best, says Nancy, these regulations are unrealistic. 'I'm from farming family. Everyone around here knows about farming. And I don't know a single farmer who'd let his land lie fallow for thirty months. They can't afford to. Most are on the edge.' Biosolids promoters also emphasize the financial attraction of their product. Chris Peot calculated that using sludge can save farmers $40,000 a year, because they don't have to spend money on artificial fertilizers, and because crops supposedly yield better and more (though sludge critics maintain that any improvement in yield is always temporary).

Nancy tells me that she and Bruce travelled around North Carolina asking farmers whether they followed the rules. The responses were shocking. 'Of the ones we talked to who were growing food crops – sweet potatoes, peanuts, melons, squash, cucumbers – not a single one had waited.' They were taken aback when she showed them her copy of the state's permit for applying biosolids, with its quotas and limits. Even if she conceded the Part 503 rules were safe, the safety is confined to paper when the rules are not followed.

'There are so many variables,' says Maureen Reilly of *SludgeWatch*. 'In practice haulers don't follow the regulations. The sludge comes out with contaminants at illegal levels. There's over-application. Animals graze it when they're not allowed to. It's an unassessable situation.' The environmental activist Abby Rockefeller, a firm believer in nutrient recycling in principle, calls biosolids 'unmonitorable, unregulatable and irremediable'.

Such views were given credence by a hard-hitting report issued by the EPA's Office of the Inspector General, a self-policing body that is mandated to investigate agency practice. Its investigation found that 'EPA regional staff in charge of overseeing the Part 503 rules numbered seven in 1998, and were down to four in 2000.' The Inspector General understood from this that the EPA considered biosolids to be a low-risk programme and devoted resources to it accordingly. But it also quoted an anonymous EPA official responsible for biosolids monitoring who classified his job as 'impossible'.

Nancy Holt thinks the Part 503 rules have more holes than her brain. They require that Class B be applied ten metres from watercourses, but take no account of possible contamination of creeks or streams by agricultural run-off. They don't address groundwater contamination. For the purposes of EPA regulation, the two creeks running behind the Holts' house are irrelevant. But that's where trouble began, in 2001, when Nancy's grandson and great-nephew were diagnosed with *staphylococcus aureus* ('staph'), a bacterial infection usually associated with dirty hospitals, and most famous for its antibiotic-resistant superbug strain MRSA. She noticed that they fell sick after playing in the creeks. Then a local dog got flesh-eating bacteria. Then someone organized a fundraiser for a couple who both had cancer, and people started taking a tally of incidents. The Cook family: three daughters with breast cancer. The Hoffmans: a mother with colon cancer, a father with prostate cancer and a thirteen-year-old son with cancer in his testicles. Five cases of brain cancer in a community of thirty-eight families. People got talking, and then they started phoning the Holt house, because Nancy had been a nurse, so she must know about disease. She started to keep records, some of which appeared in a 2004 document Nancy wrote entitled *Testimony to EPA, CDC, WEF, WERF in Alexandria, Virginia*. The list of 'health problems reported in our community associated with sludge application/exposure' include increased respiratory distress or breathing difficulties; diarrhoea (chronic during sludge applications, all ages); chronic and acute headaches (persistent after exposure to odours, relieved by leaving residence); staph infections (children covered by staph sores after playing in creeks or streams after significant rains); presumed neurotoxin sensitivity (seizures, nausea, elevated blood pressure, and rash).

Also, she sought out comrades. Helane Shields, an activist in New Hampshire, had been keeping copies of newspaper articles that reported health problems related to sludge applications. Her file was 500 pages thick. The Waste Management Institute at Cornell has gathered 350 sludge-related health complaints, and lists characteristic symptoms as: asthma, flu-like symptoms, eye irritations, lesions, immuno-deficiency, nosebleeds, burning eyes, throat or nose. In 2002, EPA microbiologist Dr David Lewis led a University of Georgia study that analysed 53 incidents where health issues had been reported near sludge sites, and found a puzzlingly high incidence of staph infections. Lewis thought chemical irritants in sludge may be causing lesions that allowed staph easy access to the bloodstream.

Doctors whom Lewis interviewed thought the cause was sludge, but they couldn't prove it. Caroline Snyder, who runs the Sierra Club Sludge Task Force, says, 'The problem is that sludge is so complex. It's a mixture of things so it's a very difficult argument. It's much easier if you can say, oh, it's benzene or some other chemical. Even three or four chemicals are difficult to analyse and in sludge there are tens of thousands. And it changes: a treatment plant will have different sludge from week to week.'

When Nancy Holt phoned her local health department to ask them to investigate why her asthmatic neighbours were forced to wear face-masks outside when sludge was being applied, she was told they could do nothing. When she noticed that she was having seizures every spring and summer, during the same periods sludge was being applied, and when she was hospitalized with neurotoxicity, she was told she could do nothing. Nothing could be proven. Everything was anecdotal.

In 2002, a panel at the National Research Council at the National Academy of Sciences released a follow-up report on sludge that is still the most authoritative document on the issue. Its conclusions were many, but the biosolids industry usually quotes this sentence from the report's Overarching Findings: 'There is no documented scientific evidence that the Part 503 rules have failed to protect public health.' Opponents quote the following sentence instead because it reads: 'However, additional scientific work is needed to reduce persistent uncertainty about the potential for adverse human health effects from exposure to biosolids.' The sentences are quoted endlessly because, in Ellen Harrison's view, 'there is a dearth of investigation in this area'. Those two sentences are the scraps that each side fights the other over.

'They say it's all anecdotal,' says Nancy, 'but they don't keep records of people's complaints. If you don't track cases or investigate them, you can honestly say there's no record.' There have been three much-reported cases involving deaths allegedly linked to sludge. In 1994, eleven-year-old Tony Behun rode his dirt bike through sludge-applied land in Pennsylvania, and died a few days afterwards of a staph infection. Behun's father said, 'We figured [the sludge] was safe. The government monitored it. Nobody ever said anything about it being hazardous. The joke was that the stuff smelled bad and grew big tomatoes.' Shayne Conner, an apparently healthy twenty-six-year-old, died less than a month after Class B biosolids were applied near his house in 1995. The third death was that of seventeen-year-old Daniel Pennock, who walked across a sludge-covered field and also died of a staph infection. His father told CBS news that he knew sludge was behind it and he would be certain of that 'until the day I die'. But nothing has reached court. The Conner family filed a wrongful death suit against Synagro, but settled. Part of the settlement, Synagro admitted, required the family to read a statement declaring that biosolids had nothing to do with their son's death. As to why Synagro chose to settle a case when it maintains that biosolids are safe, the company has yet to answer. Maureen Reilly of *SludgeWatch* has an opinion about this. 'It's the duct-tape solution. It's expensive, but it keeps everything out of the public record.'

Nancy begins to hand me documents. There are studies on the effect of lead on children's IQ levels, and on the environmental and occupational causes of cancer. There are news stories about the work of Dr Tyrone Hayes, who found that frogs were being deformed by mixtures of pesticides, even when individual pesticides were well within legal limits. There is also a transcript of a CBS story about the death of Daniel Pennock, in which EPA Deputy Administrator Paul Gilman says something he has probably regretted ever since. When he is asked if biosolids are safe, he replies with, 'I can't answer it's perfectly safe. I can't answer it's not safe.'

Nancy has the focus and energy of a zealot, because despite her calm manner and her sweet tea, she is running on fury. She talks about endocrine disruptors, chemicals found in pesticides and plasticizers, for example, which mess with human hormones. She fulminates about E. coli spinach outbreaks. She tells me that one in a hundred American children is now born with autism, and that environmental chemicals are probably the cause. She quotes the Centers for Disease Control (CDC) 'body burden' study, which tests for the presence of 128 chemicals in the human body. Lead is down; mercury is up.

Her conversation zaps like a TV remote control from lead to autism, from PCBs to prions, tiny protein particles that can cause brain diseases such as BSE. Funeral home waste in the sewage stream, she says: think about that. 'If they embalm someone with Alzheimer's, god help you. In order to kill prions you have to turn the body into carbon. That is at 1800 to 2000 degrees Fahrenheit.' She talks about how treating sludge means that of all those antibiotics in the wastewater, only the strongest and fittest survive. If we wanted to create superbugs, we couldn't do better.

Nancy's caramel accent is leisurely, but her pace is relentless. She says, 'It's so complex. Municipalities are overwhelmed,' and I'm beginning to know how they feel. 'There's all this research, but nobody's connecting the dots.' She's trying to persuade the Johns Hopkins School of Public Health to undertake a study that would link all the studies. And off she goes again, on another stream of indignant investigation.

People who promote and supply biosolids, depending on how courteous they are, tend to dismiss opponents as anything from over-emotive to hysterical. Cranks. NIMBYists. The most outspoken of all is Alan Rubin, a former EPA chemist who helped write the Part 503 rules. When we speak by phone, he tells me he enjoys 'crossing swords', and that he will talk 'from the inside of my soul'. He is a true believer in biosolids, though he thinks the name is silly. 'It got accused of being greenwash. Biosolids, sewage sludge – we all know what it is.'

He believes that the Part 503 rules take care of risk, and that if biosolids do transmit toxic elements to land, they are held firm either in the sludge or the soil. Let's take dioxin, he says. 'It's the most toxic organic chemical and the most persistent. It's everywhere. It's in our bodies.' To calculate the levels of toxins and heavy metals allowed by the EPA, Rubin and colleagues used a hypothetic individual who lives in sludge for seventy years. 'Like a farmer who works with this stuff every day, whose kids go and eat two-tenths of a gram of sludge every day for two to three years. This individual doesn't exist. But we showed that they would get ten per cent of the background exposure of dioxin.' Regular people would get 1/10,000th of it. 'God forbid that anyone should get cancer but you'd literally have to be immersed in this stuff to show you can get it from biosolids.'

Are biosolids safe? It's an impossible question to answer. As Ellen Harrison writes, 'There is no such thing as "safe". Is it safe to drive your car? Nearly all that we do entails some risk, so the question really is "is the risk acceptable?"' And risk perception is subjective. Chris Peot makes this point with a bottle of generic multi-vitamins, which contains several of the heavy metals regulated in sludge. He uses the example of selenium. 'You'd have to eat 212 pounds of our Blue Plains biosolids in a year in order to get just what your body needs.' It sounds comforting. But criticism of the biosolids programme has come from quarters that not even the fiercest sludge supporter could call emotional.

In 1975 the Chief of the EPA's Technology Branch of the Hazardous Waste Division, William Sanjour, sent a memo to his director wondering how the agency could classify sewage sludge as fertilizer, when 'industrial wastes account for twenty-five per cent of municipal sewage nationally and can be almost one hundred per cent in some localities'. Transforming this waste into fertilizer, he continued, was 'the most efficient means (short of eating the sludge) of injecting toxic substances directly into the human body'. Three years later, Sanjour wrote another memo querying why hazardous wastes used as fertilizer aren't allowed to contain cadmium, but sludge was, when nothing had yet proven it to be less hazardous. A few days afterwards, he was fired.

In a deposition to a Senate committee in 2000, Robert Swank of the EPA's Office of Research and Development, which is supposed to approve all EPA rules, said his office 'didn't think [the Part 503 rules] passed scientific muster'. He testified that the agency was assuming that toxic chemicals and pesticides leaching out of sludge would be captured by the soil, and stay put. He called this theory 'sludge magic'. 'I can tell you there was very little work actually done that looked at either threats to groundwater or threats to surface water from either toxins, pathogens, or metals that I would have called credible.' (A similar magic must be at work when Class B biosolids, which in 2000 were judged by the CDC to pose a risk to healthy, adult sewage workers, are judged risk-free even when applied to fields near young children, the elderly and the immuno-compromised.)

The next salvo was launched by Dr David Lewis. A senior EPA microbiologist, he had made his name by discovering that HIV could be transmitted in lubricants used in dental devices, research that led to new sterilization guidelines worldwide. In the early 1990s, scientists working on the Part 503 rules asked Lewis' lab for its scientific opinion. Lewis was unimpressed, and in a powerful editorial in the prestigious British journal *Nature*, complained of scientific rules pushed through by political considerations before they had the proper scientific back-up. This implied slight of the sludge programme was followed by testimony before Congress where he said it outright. To promote sludge, he testified, the EPA was using unreliable and fraudulent data. He began to act as an expert witness in sludge-related health complaints. He told reporters that 'in my opinion, the land spreading of sludge is a serious problem. We have mixed together pathogens with a wide variety of chemicals that are known to enhance the infection process. It makes people more susceptible to infections.' Taking excrement from hundreds of thousands of people, mixing it and spreading it on land is simply 'not a good idea'.

In return, the EPA cut Lewis' funding, sidelined him, forbade him to consult with other scientists, and distributed a twenty-seven-page-document that questioned his credibility, though two years later he was credible enough to be awarded the EPA's Science and Technology Achievement Award from the agency's Office of

Research & Development. The EPA's treatment of Lewis was judged discriminatory by a Department of Labor investigation, but he was still fired.

Experiences like Lewis's cause Ellen Harrison to call the wastewater industry 'intimidating'. She finds it odd that dissent is not encouraged. 'They should want to know. It's not everyone who's getting sick so what's the risk?'

Nancy Holt, who does get sick, didn't know what sludge was in the beginning. 'We called it that stinking fertilizer.' But now she knows, and the more she knows, the more her head hurts. 'We are the lab rats. I think we ought to be paying more attention to the care of humans. I know I get rabid about this, but how can we not be concerned?' She stops, after two hours of talking without a break. 'So that's where I am. Shall I serve the gumbo?'

At the end of my visit to Alexandria, Paul Carbary asked me a question. 'People don't generally stick around, so can you tell me how I can improve the presentation? We don't know how to convince people who are hostile. Do you have any ideas?' But I don't. The camps are too divided; the positions too dug in.

Responding to a *Washington Post* article querying the safety of sludge, the Water Environment Foundation declared that 'the vast consensus of the scientific community holds that land application as prescribed by EPA regulation is currently the safest science-based alternative for recycling biosolids.' I read things like this and think of a different century when engineering and science began to have total certainty, and how this confidence was expressed by Sir Joseph Bazalgette during an 1870 inquiry into the pollution of the Thames provoked by the vicar of Barking. The clergyman and 123 of his neighbours had objected to Bazalgette's practice of discharging all London's sewage into the river near Barking. One witness said he used to drink the water but now he could hardly look at it. Another said, 'I only know I have been connected to the fish business all my life time and I should think myself a madman if I brought a cargo of live fish up to Barking now.'

Bazalgette, called to the inquiry, showed himself to be as unbending as the biosolids promoters of today. The possibility that the river was being polluted was, he asserted, 'entirely imaginary and contrary to the fact'. Eight years later, the *Princess Alice* steamboat collided with a dredger near the outfall, and over 600 people died. Survivors reported that they could not swim in such noxious waters, and that they vomited copiously. The outfalls were closed twenty years later. Settlement ponds were installed to filter out the solids, one of the first attempts at sewage treatment, and the sludge was removed to the Black Deeps, at the river's mouth, safely distant. It is not recorded whether Bazalgette ever admitted he had been wrong.

Shouldn't all certainty be finite? PCBs were considered safe for decades. So was DDT. When General Motors employees who bottled liquid lead for gasoline began to die, the company called it a 'natural contaminant'.

In the absence of scientific agreement, the fate of the biosolids industry may be decided in the courts. Judges in Kentucky, California and Oregon have ruled sludge odours a public nuisance. An $18.4 million class action lawsuit has been filed against Synagro in Virginia. And in 2006, a Georgia court awarded half a million dollars to a dairy farmer who had sued for damages when thirty per cent of his cattle died after eating sludge-applied hay, ten times the normal mortality rate in dairy herds. An *Associated Press* investigation found that levels of thallium – a metal that can cause nerve damage – in the herd's milk were 120 times those allowed in drinking water. This year, Judge Anthony Alaimo, a District Judge in Georgia, found that another dairy farmer's land had been acutely contaminated by sludge. Scientific data supplied by the municipality of Augusta that claimed to prove the safety of biosolids were, the Judge declared, 'unreliable, incomplete and in some cases fudged'. Independent tests found arsenic and cadmium levels two to three times that allowed by law.

Already, food giants like Heinz and Del Monte will not accept crops grown on sludge-applied land, and the National Farmers' Union policy states that 'the current practices of spreading Class B biosolids on land surfaces [...] should be discontinued'. Alan Rubin thinks land application has peaked anyway. 'If people are complaining, and there are endless permit hearings with all the griping, municipalities and states might want to go with processes that are less trouble politically. A nice incinerator right in the wastewater treatment plant, with no need for trucks to haul it out of state.' The US's biosolids future may be already happening in Europe. Switzerland, which used to land-apply forty per cent of its sludge, banned the practice in 2003 because of the presence of PCBs and dioxins, and because farmers were already refusing the fertilizer for fear of it harming their soil. The Netherlands has banned agricultural use of sludge, and France, Germany and Sweden's national farmers' associations are against it. (The UK, however, applies over seventy per cent of its sludge to farmland and has no plans to do otherwise, because it is united with the US in soldiering and sludge.)

My stomach is stuffed with gumbo, cornbread and cobbler, and my brain is buzzing. I feel tired and spooked, perhaps due to the prospect of another six-hour drive back to Washington, DC on a dark and boring highway. It's also because once, Nancy stopped talking mid-sentence as if she had been switched off, and her husband carried on reminiscing about stock-car racing although his wife looked like she was barely living. Partly it's because Nancy told me that the holes in her brain are so rare – children are born with them, but adults rarely acquire them – that when she dies the University of Maryland will collect her brain, slice and dice it and deliver it to 620 medical centres around the world 'who want to look at that sucker', and I realized that she's probably going to die sooner not later, but I didn't have the nerve to ask when.

So I leave the Holts' bungalow in the late afternoon, wiser and none the wiser. As I pull out of the drive, I look at the field facing me. There is no visible sign of its 'guff-proof fertilizer' – as Milorganite is described – sinking slowly into the soil. There is nothing to see but an innocent field, like thousands of others, unremarkable in the darkening light.

The wall of shame, Kalyani, India – *Author*

8. OPEN DEFECATION FREE INDIA

Husband wanted; must have toilet

In a village in the eastern Indian state of Orissa, far from any city, our car is stopped by an oxen traffic jam. It's harvest time, and the oxen are on their way home. All day, I'd seen women in roadside fields forking hay with pitchforks in Thomas Hardy scenes, their bright saris and gold jewellery glittering amid yellow grass. There is no point sounding horns at the oxen, because horns are blared so often on Indian roads, they have long since lost any effect of outrage or advantage. There is nothing to do but wait, so I open the car door to get out.

'Be careful,' says Joe Madiath, who is travelling with me. 'You are about to step on the village toilet.'

Joe is a thoughtful host. He is a busy man with a huge job, as his NGO, Gram Vikas, is trying to install latrines and water supply in as many Orissan villages as possible. Nonetheless, he always remembers to point out the sights. 'That's where the toilet starts,' he says, often, pointing to the side of a road just outside a village. I don't need to specify which village, as only four per cent of the population of Orissa has a latrine, and the sanitation story is the same everywhere. Sometimes, returning in the dusk, I play a safari game. There: an old man squatting down, his buttocks exposed. Over there: a woman who has stood up suddenly at the sight of a car, and hastily pulls down her sari, but her face is resigned, not embarrassed. Here: two men walking back companionably, each holding a plastic *lota* mug that contained the water they have used to clean themselves. This standing up straight and the carrying of water vessels: these are the signs of the open defecator.

Every day, 200,000 tonnes of human faeces are deposited in India. I don't mean that they are dealt with, or sent down sewers or given any treatment or containment. These 155,000 truckloads are left in the open to be trodden on, stepped over, lived among. It is a practice known as 'open defecation', and it is done on a scale, in the words of Sulabh International, that compares to 'the entire European population sitting on their haunches from the Elbe in the east to the Pyrenees in the west'. Indians sit on their haunches from the deepest forest to the heart of the cities. They did it beside train tracks, as V.S. Naipaul recorded with disdain in 1964, and still do. The Indian journalist Chander Suta Dogra recently described an early-morning scene familiar to any train-traveller in India. 'Right in your face are scores of bare bottoms doing what they must.' Open defecation is so endemic, people do it even outside public toilets in the centre of a modern city. I see this one afternoon in the bustling city of Ahmedabad, where by early afternoon the pavement outside the toilet entrance is dotted with shit. I choose not to examine the interior.

An Indian businessman called Milon Nag tells me that when foreign clients came to visit his plastics factory near Pune, he kept a folder in the car, ready to hold up to distract his visitors from the sight of a roadside defecator. Nag is a considerate man, and the sight of people defecating in the open distressed him enough that he decided to do something about it. His accountant and sales manager became self-taught sanitation specialists. After years of research, his factory developed the most lightweight plastic latrine squatting platform on the market, now used by many big aid agencies in emergencies. 'For an Indian,' Nag says, 'open defecation is the maximum embarrassment.' That's not quite true. The maximum embarrassment for Indians is trying to defecate in traffic or a wheat field, keeping one hand free to cleanse and yet managing to cover their face with a sari, all the while watching for lurkers and lookers. Open defecation damages women most, because modesty requires them to do it under cover of darkness, leaving them vulnerable to sexual assault, snakes, disease and infection. Blocking natural body functions like urination and defecation can cause bladder and urinary tract infections and worse.

In some Indian villages, things have changed. We drive past a woman squatting and my companion says, 'Years ago, she would have stood up when a car passed, from shame. Now there's too much traffic. She can't be up and down like a yo-yo.' Also in times past, she would have covered her face. Joe calls this technique 'Cover the face, expose the base'. A woman in an Orissa village now equipped with toilets told me how things used to be. 'Anyone could see the bottom as long as they didn't know whose it was. We had to keep our dignity somehow.' How? She laughs. 'With difficulty!' (I think of a joke someone told me in Moscow: how do you use a latrine in a Russian winter? Quickly.)

The arguments against open defecation are more than aesthetic. In public health terms, the practice is incredibly dangerous. Excrement is not automatically toxic. You could probably eat your own, if you wanted to (and some people do, as do most dogs), with no harm done. The trouble with excrement is other people, and specifically other people's diseases.

Worms, for example, love to live in it. Public health experts use the term 'worm burden', a pretty way of describing an ugly reality. The world's worm burden is heavy: at any one time, about 1 billion people are carrying hookworm in their guts and expelling it in their faeces. Over a billion people have ascaris lumbricoides, better known as ringworm, which can survive in human excrement for years. In the warm environment of a human gut, an egg can produce worms half a metre long. A public health bible, *Health Aspects of Excreta and Waste-water Management*, puts the number of infections that faeces can transmit at fifty. These include salmonella, schistosomiasis and cholera. Cryptosporidiosis and campylobacter. Giardia, meningitis, shigellosis (which leads to dysentery). Hookworms, roundworms, tapeworms. Dengue, leptospirosis, hepatitis A. Typhoid, scabies and botulism.

These are all called water-related diseases, because most travel from one host to another in water. More accurately, they're shit-related diseases. Water gets contaminated by excrement, and people then drink or wash in it. Or they ingest faeces directly by the Fs, various faecal-oral contamination pathways summed up in the famous 'F-diagram' devised in 1958. Faeces can get into fluid and onto fields, fingers, flies and food. The goal of sanitation is to prevent excrement travelling from someone's anus to someone else's mouth by any of these paths. Basic, adequate sanitation is containment.

But nearly 800 million Indians are spreading possibly contagious bugs around with abandon. The scale of open defecation in India may shock because it persists in a country with a galloping GDP growth rate of eight per cent a year. Or because the Indian government has been trying to combat it for decades. Over the last twenty years, millions of latrines have been constructed throughout the country, and billions of rupees have been set aside for sanitation targets. Between 1986 and 1999, the Government of India Central Rural Sanitation Programme (CRSP) installed 9.45 million latrines nationwide, and 7.4 million more people a year gained access to sanitation. Population growth makes all targets immediately outdated (between 1990 and 1999, India's population increased by 144 million). Even so, the state's efforts increased latrine coverage by fifteen per cent. But numbers deceive. Millions of Indians may have received a new latrine, but that doesn't mean they use it.

On a road somewhere else in Orissa, Joe makes the driver stop the car. By the roadside are some modest houses, and in the yard are two small brick structures. Their walls rise shoulder high, and they have no roof. Their interiors contain grain sacks and, barely visible, a latrine pan, broken and clearly never used. Joe is scornful. 'Look! There's the Government of India's sanitation programme. There is what passes for a latrine.' He says you can find unused latrines everywhere. The country is covered with them.

All over India, in fact, those millions of government-built latrines have been turned into millions of firewood stores or goat-sheds. Surveys found latrines that were unused, misused or ignored. There are some obvious reasons why: the CRSP latrines were expensive. They were made of brick and cement. Consequently, some people found themselves with a structure that was nicer than their house. They didn't want to waste it by using it as a toilet when there was a perfectly good bush out back, and so the latrine became a temple, or an extra room. Some people had a latrine but no water supply, so they preferred to carry on defecating near a water source rather than heft the water to the new latrine. Many people didn't want to defecate anywhere near where they ate and slept – religious texts recommended defecating away from habitation – and a stroll to the defecation grounds was good for digestion. Some latrines were badly designed for the terrain, so when their pits filled too quickly, they were abandoned. Some were badly designed in the first place. They didn't have screens on the vent pipes, so flies got in. People used to the open air naturally preferred it to a dark, fetid, infested concrete box.

The lesson of the Government of India's latrine-building programme is a strange and perplexing truth: giving someone a latrine – even someone whose only other option is open defecation – doesn't mean they'll use it or maintain it. In development-speak, the government's methods were 'top-down'. They hadn't bothered to ask people what they wanted before giving it to them. At a sanitation forum, West Bengal's health minister, Dr Suryakant Mishra, illustrated this with a cartoon that showed a man balancing a toilet on his head. 'Earlier,' Dr Mishra said, 'instead of a person sitting comfortably on a toilet, the toilet was imposed on him.'

The government programme concerned itself with supply. It didn't bother looking at whether there was demand. Rural people had been defecating out of doors for ever. They didn't necessarily think there was anything wrong with it. Over the last decade, two new approaches have arisen to deal with this. Their approaches are different, but their purpose is the same: they want to make people want a toilet.

I had first met Joe Madiath in a cold hall. It was the second day of a sanitation conference, and despite the freezing temperature of the air conditioning, energy was diminishing. I was about to join the considerable numbers of delegates who had snoozed through 'Safeguarding the Health of Public Toilet Users' and 'Air Hygiene Control in Public Washrooms', when a bearded Indian man wearing a striped kurta pyjama went up to the podium.

'Good afternoon,' he said in Indian-accented English. 'My name is Joe Madiath and I work for a shit organization.' This woke me up. No one had yet used the word 'shit'. (I once attended a conference of cemetery managers, and they didn't use the word 'dead', either.) But Indians do, and especially Indians who want to sort out their country's execrable sanitation situation. They have no time for verbal niceties. Even so, Joe apologized for his 'unparliamentary language.

But these words are used everywhere every day so allow me to use them.' Then he told a story.

Akbar was a great Mughal emperor. One day, Akbar asked his favourite minister Birubar what the most compelling need of a human being was. When Birubar said, 'Sire, to shit and piss,' the emperor was offended and dismissed him. A few days later Akbar was on his boat in the River Ganges and he came to need the toilet. But Birubar had made sure there were no toilets on the boat, and Akbar became desperate, until Birubar sent him to a secret place where there was a toilet. Akbar came back and said, 'You are right. The most compelling need of the human is to shit and piss and it is also something that gives you the greatest amount of relief.' Birubar kept his job.

Akbar's epiphany came late to Joe. In 1971, equipped with a degree in English, he went to work with refugees who had fled from the newly created Bangladesh. Later that year, a cyclone hit Orissa, bringing a poor state already on its knees flat to the ground. With a few friends, Joe headed for Orissa to see what he could do. He meant to stay for a year and hasn't left since.

Today he runs Gram Vikas ('Village Development'), whose headquarters consists of a rural wooded campus in beautiful scenery five hours from Bhubaneswar, Orissa's capital. There are no mobile phone signals, there are bicycles to borrow, and the campus canteen makes excellent Indian chow mein for breakfast, at least when the monkeys aren't baiting the cooks. The place runs efficiently and confidently. Gram Vikas has won major international awards, like the 2006 Kyoto World Water Grand Prize, a big deal in the watsan world.

Yet its mission didn't start with toilets. The young Joe Madiath began by lobbying for land reform, because people were moving to the cities when life on the land got too hard, and the land was suffering. Eventually, his focus shifted to biogas, which also made poor farmers' lives easier. Over fifty thousand cow-dung digesters were installed over ten years. Still it wasn't enough. Gram Vikas conducted field surveys to find the most acute problem stifling village development. They concluded that it was poor health linked to excrement. More than eighty per cent of diseases were caused by contaminated water supplies. Sixty per cent of women had skin and gynaecological diseases because they were bathing in faeces-contaminated water.

Orissa is not a water-starved state. Each village has a bathing pond, but this is also the animal trough, wash-tub and toilet. The most pressing problem, Joe realized, was endemic open defecation. Shit was in the drinking water, on food, on the roads, on the bottom of flip-flops, on bare feet. Diarrhoea was a weekly event for most people, and many children died from it. The obvious solution was to build latrines.

If Joe had been an engineer, or a bureaucrat in the CRSP programme, he would have done only that. Orissa would have had a lot more goat-sheds. But it seemed obvious to him that imposing toilets on some people wouldn't work unless everybody had one. It had to be total, and it had to be consensual. 'I knew there was no point in doing anything if eight families had toilets and ten shat in abandon.' The ten open-defecating families would continue to contaminate the water, and the living environment, of everyone else. If sanitation were supplied, it would have to be to one hundred per cent of families.

These realizations, the result of observation, investigation and common sense, were actually significant policy shifts from traditional development practices. Until recently, sanitation programmes always focused on the household. The family unit was the one to be persuaded to change its habits and to build the toilet. But that was treating sanitation like water, when the two are nothing like. A family can be the only household to install a water pipe in a village, and it can use clean water without harming anyone else or being harmed by anyone else. But it only takes one family without a latrine to pollute all common areas and drinking water. The irony of defecation is that it is a solitary business yet its repercussions are plural and public.

Joe thought the solution was two-fold: all families in a given village would have to agree to build a toilet and bathroom, and all the families would have to agree to pay for it. He calls this method 'one hundred per cent sanitation'. This was also radical. It has been a standard in development thinking that poor people need subsidies. They aren't supposed to be able to pay for things without help. In fact, poor people have money, but their money is busy. They prioritize. A toilet is rarely considered a priority when there is food to buy and school fees to pay, even when the lack of a latrine contaminates that food and makes children too ill to go to school.

Gram Vikas' job was to persuade people otherwise. Joe could see that there was no point supplying latrines unless he also provided water to cleanse with. That had been the government's mistake. He tells the story of a daughter-in-law in a village where toilets had been installed. 'We assumed that people would bring the water needed for the toilets from nearby hand-pumps. But this job was relegated to the wife, daughter or daughter-in-law. One day, a daughter "accidentally" dropped a small stone in the U-trap of the toilet, making it unusable. Pretty soon, similar accidents were taking place in other households, and it wasn't long before people went back to open fields for their sanitation needs. Our lesson was clear. If water-based toilets are to function, we need to make sure running water is available.' The lesson was also that humans are complicated, and that sanitation is never only about bricks and latrine pans.

Gram Vikas chose to pilot its method in the village of Samiapalli. It's now the Gram Vikas success story, where visitors are always taken. But it used to be an unhappy place. There were several castes, including ostracized Dalits, who rarely got on. Alcoholism was endemic, as was domestic violence. Women had no say and kept their faces covered whenever they left the house. No-one had a latrine and everybody did open defecation. All this had to be changed. The weapons were patience, wits and bribery.

On the face of it, Gram Vikas was arriving with a gift horse. They were offering to contribute to building a toilet and bathroom for each family. They would also provide twenty-four-hour running water by digging bore wells or channelling nearby springs, and they would build a tower to store the water in. But they were also asking for money. Under the Gram Vikas model, each family had to contribute 1000 rupees ($25) to a common fund. Gram Vikas would contribute the big costs – cement, doors, latrine pans – and the fund would pay for the rest and then for upkeep and repairs. Families who couldn't afford to pay cash could pay with labour or in kind. Still, when the daily labouring wage was less than a dollar a day, Gram Vikas was asking for the moon.

The persuaders had to craft their message carefully. There was little point telling villagers they were risking their health, because health messages rarely have an impact. I call this the 'doctors who smoke' theory. Doctors know the harm smoking does, but they smoke. Reason rarely persuades people to change behaviour. Research in Benin into why people want a latrine found that their principal reasons were to avoid embarrassment when visitors came; to make the house complete; not to have the chore of walking to get water; and to feel royal. Improved health never came into it. Dr Val Curtis of the London School of Hygiene calls hectoring health messages the 'doctors, disease and diarrhoea' approach. It never works, and it's not culturally relative. When Curtis did covert surveillance of hand-washing rates in her own institute, she found that male students – who are, let's remember, students at the London School of Hygiene, and well-schooled in disease transmission – failed to wash their hands after the toilet in sixty per cent of cases. Only when the surveillance was announced by email did hand-washing rates leap to eighty to ninety per cent. Curtis likes to refer to the success of soap manufacturers, who got soap into nearly every household in the early twentieth century, not by promoting its health benefits, but by telling consumers it would make them smell better, be more attractive, be sexier. Soap companies understood that they should target their product to wants, not needs.

In Samiapalli, Gram Vikas had to create a want by bribing people with a need. They focused on women. They were the ones who fetched water from the pumps. In Samiapalli the pumps weren't far, but water is heavy. Gram Vikas told them they could have a tap in their kitchen if they'd build a latrine and bathing room too. The bathing room was on Joe's insistence, because his catchphrase is 'building dignity through toilets'. He'd seen enough women trying to wash themselves under their saris at the bathing ponds, and how that brought scabies and vaginal infections by the score. Villagers had to agree to build the latrine and bathing rooms up to roof level, and then Gram Vikas would bring water. Sojan Thomas, Gram Vikas' sanitation director, calls this, smiling under his moustache, 'a by-force business'.

It took 162 meetings, and 2 years of talking. Gram Vikas appealed to their prejudices. Sojan gives an example. 'Often richer people say, "Poor people don't need toilets, they're dirty people. Let's carry on the programme without them." And we say to them, "These people are still shitting by the pond. A fly that has touched their shit is not going to distinguish between Brahmin and Dalit food. If you have toilets and they don't, that means that your food is definitely being contaminated by lower-class shit." We appeal to their ego and it usually works.'

After the latrine-building project was complete, people realized that their toilets were better than their houses. So they applied for housing loans, with Gram Vikas as a guarantor, and now they have concrete *pukka* houses that don't blow away in a cyclone. (When anthropologists complain that Gram Vikas is not preserving 'authentic' thatched roof houses, Joe suggests they try living in an authentic thatched house in a monsoon.) Women got enough confidence to start speaking in meetings, then to tie wife-beating husbands to lamp-posts, then to set up self-help groups. With the time saved from fetching water, they can sow fields of peanuts or cultivate fish. Children's attendance rates at school have shot up, because they are sick less often and because there are fewer chores to do around the house. Before toilets, ten per cent of girl children went to school. Now it's eighty per cent. But the biggest thing they have got now, the village leader tells me with a big smile, is pride. 'We are better than higher caste places! People from higher caste villages want to come and live here!'

A toilet for Gram Vikas is never only a toilet. Joe calls sanitation 'the entry point'. It's the most difficult entry point and the one that people are least likely to agree on, but once they do, anything is possible. A toilet can change society in unexpected ways. For example, it can improve your marriage prospects.

At Bahalpur, we arrive unannounced. A woman runs up to give me a bunch of flowers, nonetheless, and a meeting is swiftly convened on a veranda. I have learned by now to recognize a Gram Vikas village. From afar, by the yellow water tower; and from nearby by the Gram Vikas slogans and posters painted on the walls. Common to all the villages I visit is a poster that shows a middle-aged woman in the foreground, and in the background a young woman entering a clean-looking latrine. Translated from Oriya, Orissa's state language, it reads: 'I will give my daughter in marriage only to a village with a toilet and a bathroom.' I'd taken this as the propaganda it looked like, but in Bahalpur I changed my mind.

On the veranda, men with towels draped round their shoulders sit on one side and women on the other. But the women are unveiled and they are chatty. They crack jokes. Joe is astonished, because he hasn't seen this before, and because in the beginning, the women wouldn't even come to meetings. I ask the women whether the poster is true. 'Oh yes, our girls don't want to marry into families from Malad and Borda, because they don't have toilets.' In fact, a girl has just left her toiletless in-laws to return to the village, and she arrived home the day before. The girl, whose name is Gilli, is fetched to be questioned, and sits at the back of the group, her eyes cast down. She looks mortified. She's not supposed to speak to strangers, because she has been married less than a year and has no children. But the women give dispensation, and Gilli says, 'It was very difficult. The drinking water there is from the pond. They shit on one side and take their bath on the other.' She's trying now to pressure her parents to persuade her husband's family to build a toilet, but they may also send her back, toiletless. The conversation moves on, and shortly afterwards I see Gilli halfway up the hill, walking away as fast as possible.

But girls have to be married, so the villagers are trying to persuade their neighbours to sign up for the Gram Vikas project. This is sanitation contagion. The more villages that join, the easier it gets. Now, any village that is interested can send representatives to stay in a Gram Vikas village to experience a toileted lifestyle. It is one-way traffic. 'We do go to other villages,' a towel-wearing man says, 'but it's very hard for us now to visit villages where they defecate in the open. We stay for dinner but we'll always come back.' When relatives come to stay, they are told to use the latrine or else.

Or else what? Villagers have learned about compliance because they had to. In the early days of the project, with the latrines built and the water supplied, people were still going for open defecation. They were too used to 'going out there in the open with the wind in their sails', in Joe's words. (I get fond of Joe's words, especially when he later comes to London and we go to a posh Chinese restaurant where he orders crab, then says loudly, 'I love crab! Crab and crap!' and does a bellow of unparliamentary laughter.)

The village council already had the practice of fining people who transgressed village rules in some way. It was easy to set up a new system of open defecation fines. Defecating in the open would cost fifty-one rupees (one rupee is added to the round number for auspiciousness). The person reporting the offender kept half the fine. The rest went to a village fund. Joe likes to say that the toilet can build livelihoods. With this fining system, a new livelihood of Toilet Spy came into being. Some people gave up their day labouring because reporting people was more lucrative. The women giggle about it now. 'For the first three months people continued to go outside. But we could spy on them because they always carried an aluminium vessel of water. Then after they were caught they would still try to sneak outside and hide a bottle under their arm.' The repeat offenders were caught too, until open defecation was finally banished. No-one would dream of doing it now.

Some might find such a system too coercive. Perhaps it pits villagers against each other. Perhaps it's not democratic enough. It's all relative: in Sierra Leone, one NGO suggested bringing in the army to get people to use latrines. Anyway, fines were already imposed for transgressions. In fact, all that has happened is that morals have entered the toilet room. There is now a right and a wrong defecation behaviour, and toilet habits have become a communal concern. Sanitation projects often falter, says the sanitation specialist Pete Kolsky, because of a mentality that considers that 'my crap is your problem'. In Gram Vikas villages, everyone has acknowledged that it's everyone's problem. This is a very big deal.

The Gram Vikas method is effective, but it is slow. Today 361 villages have one hundred per cent sanitation, but there are 50,000 villages in Orissa. At present rates it would take Gram Vikas centuries to reach them all. This is the burning question for sanitation experts.

There are good projects everywhere, but how can they be spread? In development language, how can they go 'to scale'? Kolsky talks about 'the false question. You look at a jewel of a development project and say, why can't we multiply this by a thousand? Because you can't. You've got to have something that can be done by Joe Shmo or an ordinary bureaucracy.' Joe Madiath admits that his two great problems are finding human beings who want to live in villages for the two or three years that projects usually take, and finding funds. 'Ours is a difficult model,' he says. 'I always face an inquisition from the government because we insist on one hundred per cent compliance. They say it's too much. They are content to say that whoever wants can have a toilet and never mind the rest.' But he's stubborn. 'We believe that if it can be done in Orissa, in the poorest state in India, it can be done anywhere in the world.'

The Gram Vikas model provides subsidy because Joe believes in quality. Technology for the poor, he says, should not be poor technology. But in the years that Gram Vikas has been sweating away in Orissa, a conceptual shift has taken place in government and in sanitation thinking. Subsidies, an essential of India's latrine-building programme, have fallen from favour.

India's efforts to eradicate open defecation are directed from New Delhi by the Rajiv Gandhi National Drinking Water Mission, part of the Rural Development Ministry. Its offices are on the eleventh floor of a tower-block whose lifts only stop at even-numbered floors. I go to the twelfth, assuming I can walk down, but this floor houses the Indian Air Force's Adventure Corps, as is revealed by several photographs on the walls of dashing Indian men with black moustaches and Himalayan backdrops, and by a locked steel gate blocking the stairway. Presumably it's more adventurous if in an emergency the Air Force officers leave by the window.

I arrive at the office of Sanjay Kumar Rakesh, director of the

RGNDWM (an initialism that is actually used) late and in a fluster, and it's an appropriate start, because in a twenty-minute meeting Rakesh answers half a dozen phone calls, two emails, and deals twice with his assistant, all the while failing to introduce or explain the silent man sitting next to me. In between, he finds time to tell me that things have changed radically. The CRSP is dead; a new programme called the Total Sanitation Campaign (TSC) has risen in its stead. The TSC has things in common with the Gram Vikas model. It still dispenses subsidies, though fewer than before. Now, only poor families can get one, and only $11, a third of the old subsidy. But a more important change was to recognize the power of persuasion, as demonstrated in a remarkable regional sanitation programme in the Bengali district of Midnapur. Its Intensive Sanitation Campaign, carried out by the local Ramakrishna Mission and UNICEF, abolished the subsidy-heavy model of the TSC in favour of education and persuasion. The software was backed up by hardware made easily available in a network of Rural Sanitary Marts, local shops selling sanitary equipment at reasonable prices. It was a big success, and helped to inspire a change in policy. Before, the success of sanitation programmes had been judged by compiling statistics about coverage. In short, counting toilets. The Intensive Sanitation Campaign, and other similar projects throughout India, tried a new approach. They counted, instead, how many people were still going for open defecation. A new acronym – ODF, for Open Defecation Free – was created to sum up the new goal.

The TSC also aims to achieve an ODF India, though Rakesh is still counting toilets. He rattles out some figures. Coverage is increasing by 7.5 per cent a year. Extensive monitoring. I ask to see a monitoring report and he tells me to check the website. (I do, and the links don't work.) I ask about the Millennium Development Goals and he shrugs. 'We're not concerned with those. We have already met them. We have to reach fifty-five per cent coverage by 2015 and we're already at forty-five per cent.' He seems bored. He says, 'Everything is on our website. Is there anything else?'

It is a frustrating interview, but it shouldn't detract from the fact that the TSC is an improvement, or that Rakesh's boss, Rural Development Minister Raghuvansh Prasad Singh, has said that 'a toilet or the lack of it is the indicator of a country's health, not the GDP or Sensex [the Bombay stock exchange]'. And he has put money behind this conviction. The TSC is lavishly funded. With a budget of $810 million, it is twenty times bigger than the Bangladesh sanitation budget, though the number of India's toiletless is only ten times that of Bangladesh's. The Minister has creativity as well as cash. In 2003, the Nirmal Gram Puruskar (Clean Village) prize was launched, awarded to villages judged to have one hundred per cent toilet coverage, one hundred per cent school toilet coverage and to be open defecation free. In the prize's first year there were 35 winners out of 70 entrants. The following year there were 770 entrants and in 2007, 10,000 villages applied.

The awards are handed out by the President of India and get serious media coverage. Other efforts by Minister Singh have been more controversial, such as a letter he wrote in 2005 to all chief ministers demanding that they pass a law requiring anyone running for local office to possess a latrine. No toilet, no election.

TSC still had problems. Not all regional officials wanted to abandon subsidies, because doling out subsidies makes politicians popular with voters. So some states have diverted funds to keep their subsidy levels high, with predictable consequences. Researchers found half the latrines unused in some TSC project areas. Something was still not working. Meanwhile, over the border in Bangladesh, something was.

On a day in 1999, an Indian agricultural scientist called Kamal Kar arrived in the Bangladeshi village of Mosmoil. He was there as a consultant for WaterAid, which had asked him to assess whether the organization's subsidy approach and latrine-building programme was working. WaterAid couldn't understand why its Bangladeshi branch had been building latrines for years, but forty per cent of the country's illnesses were still the excrement-related kind. Kar thought WaterAid was asking the wrong question. 'Let's not talk about subsidy,' he told his employers. 'Let's find out instead why people are shitting in the bush.'

I meet Kar in his spacious apartment in Salt Lake City, an affluent suburb of Kolkata. He has an intense energy which shows through a ceaseless jiggling of his legs and speech at the speed of a machine gun. He is blunt. 'You can't be a doctor and be scared of blood, and you can't work in sanitation and be scared of shit. Anyway, no-one understands you when you say sanitation.'

At Mosmoil, Kar used techniques from a discipline called 'participatory rural appraisal'. (A lay-person may ignorantly translate this as 'asking people you're trying to help what they think'.) This usually involves a walk through the village – 'a transect walk' – and asking locals to draw a map of their surroundings. Kar did this, but the transect walk, once it had passed through the nice parts of the villages, carried on to the areas used for open defecation. On the map, once the houses had been chalked in, villagers were asked to indicate where they usually went to defecate. As Kar explained in a how-to guide to the method, 'It is important to stop in the areas of open defecation and spend quite a bit of time there asking questions and making other calculations while inhaling the unpleasant smell and taking in the unpleasant sight of large-scale open defecation. If people try to move you on, insist on staying there despite their embarrassment. Experiencing the disgusting sight and smell in this new way, accompanied by a visitor to the community, is a key factor which triggers mobilization.'

The calculations involved villagers doing their sums. They were asked to reckon how much excrement was being left in the open. 'The accumulated volume of faeces,' Kar wrote, 'is reckoned in units that can immediately be visualized by the community – cart-loads, truck-loads, boat-loads. There is much amusement as people reckon up which family contributed the most shit to the pile that morning. But as the exercise goes on, the amusement turns to anxiety. People are horrified by the sheer quantity of excrement left in their village: "120,000 tons of shit is being dumped here every year? Where the hell does it all go?"'

The answer, as the villagers of Mosmoil figured out for themselves, is 'into their bathing ponds and rivers; and from there onto their clothes, their plates and cups, their hands and mouths. Onto the udders of their cows and into their milk. Onto the feet and hooves of their livestock, dogs and chickens, and onto the flies that carry it straight to their food.' Eventually, the villagers of Mosmoil calculated that they were eating ten grams of each other's faecal matter a day. At this point, the brilliant core of Kar's method is revealed. The brilliant core is disgust.

Disgust is probably the least studied of all human emotions. It has been called the forgotten emotion of psychiatry. Opinions still vary as to its composition, function and genesis. William Ian Miller, in *Anatomy of Disgust*, one of the few serious books on the topic, sets out the two main theories. Biologists, he writes, think disgust is innate. What is disgusting is usually what is bad for you.

This was given credence by a huge online poll carried out by Dr Val Curtis. Participants were asked to choose one of two similar pictures. In all cases, the picture judged to be more disgusting – the greeny-brown soup over the blue gloop; the worms, not the caterpillars – showed something that could carry disease. Greeny-brown looks like body fluids, which can be dangerous. People have worm burdens not caterpillar burdens. Humans experience disgust, Curtis theorizes, because it keeps them alive. Anyone who doesn't find excrement disgusting risks contracting diarrhoeal diseases from getting too close to it. 'If you've got an innate capability to avoid things that are going to eat you from the outside,' she tells me, 'then probably you've got an innate ability to avoid things that can eat you from inside. Parasites are bad for you because they make you sick and die and you won't reproduce. Also they make you unattractive and you won't reproduce.' Our disgust is also visceral. When volunteers were blasted with skatole, the stinkiest compound in faeces, none could stand it for more than five minutes, and all expressed physical signs of disgust such as facial expressions and pallor. This needn't trouble Curtis' theory: what is more visceral than the need to survive? Another study found that when presented with a selection of dirty, unlabelled diapers (one of which belonged to their own child), mothers regularly ranked their own babies' faeces as less disgusting than others. Parenting, desirable for the survival of the species, would be compromised if mothers were distracted by disgust.

Anthropologists think disgust is learned. They point to small children who show no disgust at dirt or faeces until they are educated otherwise. The anthropologist Mary Douglas concluded that something is dirty because it is out of place. Soil in the garden is fine; soil on a plate is not. Disgust becomes a way of ordering a society, of creating a hierarchy of what is safe and what is acceptable. It also becomes a way of distancing intellectual humans from their embarrassingly animal origins. John Berger, in an essay for *Harper's Magazine* about cleaning his outhouse, concluded that 'what makes shit such a universal joke is that it's an unmistakable reminder of our duality, of our soiled nature and of our will to glory. It is the ultimate *lèse-majesté*.'

None of the theories seemed to hold true in Mosmoil. For reasons of convenience and habit – they had no latrines, so they had to go somewhere – the villagers had suppressed disgust. Kar thought the only way they would change is if they did it themselves. It is difficult, he believes, to break the entrenched habit of development professionals; to resist being the omniscient outsider coming into the village and dispensing instruction and free latrines. Yet it was crucial. Any awareness had to be revelation, not instruction. From within, not top-down.

In Mosmoil, after the faecal calculation, people started vomiting from the shock. Then Kar did something more shocking still. He left them to it, or threatened to. 'I said, "Carry on what you're doing. Your forefathers did it; you can do it. Good-bye."' The story as Kar tells it is suspiciously dramatic, but enough reports have been written on the Bangladesh programme that I believe him. Immediately, he says, the villagers were fired up with shame and disgust and determination. Children ran off to start digging latrine pits on the spot. The villagers swore that within two months 'not a single fellow would still be shitting in the open'. All this took place without a penny of subsidy being dispensed. No latrines had been supplied, no technical advice. In the how-to guide, Kar says that once disgust has been triggered, villagers may say they can't afford a pit latrine. At that point, the facilitator can suggest a simple, low-cost design, emphasizing that it was created by poor people. Kar wanted to shift the focus away from hardware. It didn't matter, he believed, if latrines were temporary. People would upgrade if they needed to. Once they'd seen the light of disgust they would do whatever was necessary.

Kar went back to his hotel after a day in Mosmoil with the makings of a new methodology that apparently worked. He came to call it Community-Led Total Sanitation, or CLTS. Community-led, because it was not about outsiders imposing things on insiders. Total Sanitation, because it kept the one hundred per cent requirement of the Gram Vikas and TSC models. In Bangladesh, where WaterAid had good local partners, and where the population density made people receptive to a private latrine, its success spread fast.

*

In India, where development experts were sick of seeing their expensively provided latrines standing unused, and tired of seeing those 'bare bottoms doing what they must', CLTS looked like a ray of hope. Kar initially refused to try it in his home country. India's huge bureaucracy meant that there were too many government meddlers to interfere with things at village level. Bangladesh had nothing and its people had nothing. It was easier to persuade people to do things for themselves. In India, too much money was still being thrown at sanitation. People had got dependent on subsidy. They saw the next village getting subsidized latrines, and preferred to wait their turn for hand-outs rather than build their own. Kar told his countrymen that CLTS would never work in India. Eventually, thanks to persistent persuasion from the Water and Sanitation Program's Delhi office, he changed his mind.

Kar invites me to Dharamsala, where he's running a CLTS workshop. Dharamsala, the involuntary home of the Dalai Lama, is one of the most popular tourist spots in the Himalayan state of Himachal Pradesh. Tourists come here to see Tibetans or mountains, but it's lucky they don't come to see toilets because they wouldn't find many. Himachal Pradesh's sanitation statistics are woeful. Eighty per cent of people do open defecation. The state government has therefore decided to initiate CLTS state-wide. HP is a test case for CLTS, and the Dharamsala workshop will train the people who will make it a success.

The workshop is held in an unheated hall. The temperature outside is near freezing, but I'm the only one feeling the cold. Everyone else is making up for heating with passion, energy and woollen Himalayan shawls. The attendees are a mix: hygiene workers; women from self-help groups; ordinary villagers. They have spent the previous day learning the basics of CLTS. Don't tell, always ask. Awaken disgust. Use crude language. Never, ever promise anything.

They are enthusiastic, itching to try it out, so a village is assigned to each of five groups, and off they go.

Kar and I set off in pursuit, but logistics intervene. In the first village, there's nobody around to test the method on, and it takes us too long to get to the second. So I arrive in a small settlement outside Dharamsala to find villagers who have already been 'triggered'. On the ground by the village office, there are still traces of the yellow chalk outline of the map, and two handsome men with cheekbones as sharp as the mountain-tops are standing on a path nearby, looking somewhat bemused to a backdrop of Himalayas.

In one man's hand is the paper on which villagers have done their faecal calculations. 'It was ten truck-loads of shit a year,' he says, and holds up a sheet of paper with figures written on it. 274 kg per day. 8220 kg per month. 98,640 kg per year. That sounds like a lot, I say. They nod. 'People didn't have any understanding until the outsiders came, and that's when we realized, what the hell are we doing?' I ask them what changed their mind, and the older man says, 'When this guy picked up a hair and touched some shit and dipped it in a glass of water and asked if we would drink it now – that was really shocking. Also they brought some food and we saw that the flies were flying between the food and the shit. That was horrifying. It was only a short hair but flies have six legs.'

He's referring to my favourite part of the triggering process. Facilitators who do the transect walk can, if they choose, surreptitiously remove some faeces from the open defecation areas and bring it back to the meeting point. There, they will place it next to a plate of food – a chapatti bread or rice will do – and say nothing. Flies will soon arrive and fly between the two, and eventually someone will notice. It is a strikingly effective way of communicating contamination. There are other techniques, often devised by the 'natural leaders' who sometimes emerge during the triggering. One

Bangladeshi villager devised the phrase 'a fly is deadlier than a thousand tigers,' now a slogan nationwide.

That's home-grown communication. CLTS also breeds home-grown innovation. On his laptop, Kar has dozens of photos of latrine-building ingenuity, where villagers have used what's readily available in the place of expensive equipment. There's a latrine pan made from a cement pipe cut in half; pipes made from plastic soft-drink bottles; walls of tin or tarpaulin. Once a village has been ignited, CLTS facilitators give some training in building a safe latrine pit, siting it a safe distance from water courses, but beyond that, it's up to the villagers. This approach infuriates some engineers and public health specialists, who worry about groundwater contamination from poorly sited latrines, or who say that latrine-building is a specialist affair. Kar has no patience with this.

Sanitation experts talk about a 'sanitation ladder', where people begin with the poorest safe containment option and move on up to the flush toilet. The latrine gets upgraded like a mobile phone. For Kar, these low-tech, flexibly designed latrines are a new first step on the ladder. He is convinced that people will maintain and upgrade them because self-motivation is more sustainable than anything else. He proves this point by taking me to a warehouse in a city outside Kolkata, where an innovative local government chairman is trying to make his municipality the only urban area in India where CLTS is working. In Kar's terminology, 'the only city in India that can claim that no bloody fellow is defecating anywhere'.

Kalyani, two hours west of Kolkata, has fifty-two slums in its jurisdiction. This is no black hole: these slums are classed as 'peri-urban'. They are spacious, with trees and banana plantations, and good-sized shacks. But they are still illegally squatted, and infrastructural services are minimal. Toilets are either non-existent or unused.

In the town-hall yard, Kar opens a garage door and points. There is a pile of Nag-Magic slabs, manufactured by Milon Nag in his Pune plastics factory, and costing an affordable 120 rupees ($3) each. Kar ordered 1000 and expected them to be snapped up after CLTS had been done in Kalyani's slums. He came back to find that only 21 had been bought. 'No-one wanted them. They wanted something better.' As Gram Vikas knew, toiletlessness is not always about poverty. Kar tells the tale of a wealthy farmer in Haryana. 'He would take his whole family out in the morning for open defecation, in his nice Maruti car. They would shit in the bush and come back in the car. It's the whole mental thing!'

Our visit is hosted by two female doctors. Dr Shibani Goswami works for Kolkata's Urban Services for the Poor (KUSP), which is attempting to implant sanitation and good hygiene in the greater Kolkata area. KUSP has a substantial budget, so it gives out generous subsidies – 9900 rupees ($200) for a latrine. Kar scorns it. But he likes Dr Goswami, because she thinks 9900-rupee subsidies never work either. Dr Kasturi Bakshi, meanwhile, works for the municipality. They seat us at a long meeting table, on chairs whose backs are covered by towels to accommodate sweaty councillors. On the wall, there is a series of red, yellow and green cards, each with a passport photo, a name, and some figures. This is the Kalyani wall of shame.

CLTS relies on manipulating emotions to work. Disgust is the first, but then come shame and pride. Villagers are encouraged to boast about their ODF status. They put up posters – 'No-one defecates in the open in this village!' – and crow to their relatives. It's supposed to foster competition and usually does. The wall is designed to achieve a similar result. Each card represents a local councillor. Each local councillor is responsible for so many slums, and each colour represents how many slums are now open-defecation-free. It's fabulous and, says Dr Bakshi, it works brilliantly. 'When councillors come in here, the first thing they do is check the board. "Oh God, I'm still in the red." There is no need to say anything.'

We take the municipality's ambulance to Vidyasagar Colony in Ward 5, one of the first slums to be triggered. The colony has wide sandy streets and walls decorated with cow patties slapped against them to dry. Our first stop is at the residence of a widow. Sandhya Barui is sixty years old. She came to West Bengal from Bangladesh thirty years ago, and she's been living in this slum ever since. Her place looks quite nice. She has a decent shack, adequate space, and a sturdy latrine that she built herself. Even the pit. She's a frail, skinny woman, but she's stronger than she looks.

It's not a sophisticated latrine – a cheap plastic pan, sheltered by walls and a roof of bamboo, banana leaves and plastic – but it is swept and obviously used and prized. Behind Barui's shack is her old toilet, a field of bananas whose leaves never grew tall enough to hide her from peering children. 'They would come and look at you,' says the widow, 'and laugh at you while you squatted. It was sinful.' Also, someone in her family got diarrhoea at least once a month. She spent 100 rupees ($2.50) a month on medicine. 'To us, even if we lose five paise [less than a tenth of a cent], that's a lot of money.' Nonetheless, she spent 700 rupees on her latrine and she thinks it has been worth every paisa.

The widow's pleasure in her latrine is infectious, and I understand why she's a standard stop on the CLTS tour. But her latrine still constitutes a risk because her shack is technically illegal. In fact, when the slum was declared open defecation free, the mayor faced a problem. He wanted to reward the colony, but he couldn't give them the title to their illegally occupied land, as the slum-dwellers wanted. He couldn't give them anything, either, that might be construed as conferring land-rights, like electricity. So he gave them three solar street-lights, now the only street illumination in the whole slum.

We stand in front of a lamp-post, facing a large map tacked to a shack that displays which residents have built latrines and where. Our guides are a quartet of young local men. They are the 'natural leaders' – the most persuasive and articulate villagers – who arose out of the triggering process. Natural leadership is now their job. Their task is to trigger CLTS in nearby slums, and for each house triggered, they are paid 100 rupees. They tell us that they've just completed a village of thirty-seven houses. Kar is practically jumping up and down with pleasure. '3700 rupees! Oh my God! That's big money!'

The young men no doubt deserve the reward. They have got results. But triggering is a delicate process, and it can go very wrong. I see this on the third day of the Dharamsala workshop, when the five groups have to report back on the success of their triggering the day before. Kar gives the volunteers his categories of success. Proper ignition is 'a matchbox in a gas station'. Partial ignition and some cause for hope is 'fire under ashes', because something is smouldering. Next comes, simply, 'hope'. The worst is a damp matchbox. Each team stands up and delivers its report to the hall. Kar mostly keeps his counsel, even when Group 2, for example, delivers a soaking wet matchbox. People in their assigned village hadn't been warned and hardly anyone was around. The trainees had used a crude word for 'shit', and women had got embarrassed and covered their faces. The faecal calculation was done all right, but then the trainees asked villagers how much faecal contamination they ingested. Well, none, said the villagers.

The young trainee presenting the report looks embarrassed. 'Then we said, how much contamination do the dogs bring in and they said they didn't have dogs, but we could hear them in the village. When we asked about the source of water, they said it came from a place with no contamination. When we asked about flies, they said, "We have no flies."'

By this point most of the workshop is laughing, and not all of it is in sympathy. Other groups do better. They get the required disgust and promises. Some of their members may even become natural leaders. Some already have. At the end of the day, a thin-faced man wearing a baseball cap approaches me. He is bursting to tell his story. His name is Mansingh Kapoor, and he works for the Natural Environment and Health Association in Palampur. He is an educated man whose job requires him to know – and care – about hygiene. He says, 'I want to tell you that yesterday I was doing open defecation but today I have stopped.' He has had a latrine for five years and never used it because 'we have had the habit of open defecation for one hundred years. It's nice to go out and sit by a river in nature'. But now he's had his CLTS epiphany. He will start using his latrine on Monday. And, if Kar's theory of contagion works, he will convince other people to do something about it.

The contagion theory holds that if CLTS doesn't work in one village, then you leave and start on another one nearby. A film about CLTS made by the World Bank Water and Sanitation Program (WSP) shows a young bride sitting in a room. Her sari is red; her attitude is demure. Two men enter and hand her a latrine pan. 'There have been instances,' the narrator explains, 'where health motivators were seen gifting latrine slabs to brides. This ensured the bride not only continued to practise safe sanitation in her new village, but also set an example for her new neighbours.' If you can't find a husband with a toilet, take one with you.

CLTS is seductive. The results, when they occur, are visible and relatively immediate. Not everyone is persuaded. Monitoring so far has been scanty. Indian journalists who followed up villages that had won a Nirmal Gram Open Defecation Free award returned a year later to find dirty, unused latrines and open defecation still going on.

Also, CLTS might flounder after the first low-tech pits collapse. Kar answers that with an anecdote about a man at a Bangladeshi meeting, where engineers presented their new low-cost latrine which cost 300 taka ($4) and had taken years to develop. 'Then this old man got up and said he'd built one for ten taka. They said, '"But ours last for ten years!" and the man laughed. "I change my roof every two years. Why not my toilet?"'

Nonetheless CLTS has some powerful supporters. Soma Ghosh Moulik, a senior sanitation specialist at WSP India, is a firm believer. 'It has to involve local government,' she says, and if that can be done, the possibilities are huge. I ask her when she thinks open defecation will be eradicated, though I don't put it quite that nicely. I say, 'When will we see the last bum on the railway tracks?' She doesn't hesitate. 'If TSC goes the way it's going, focusing on [building] toilets, then 2024. If TSC moves towards CLTS, 2010.'

In the newly toileted slum of Vidyanagar in Kalyani, I notice a woman with bottle-top glasses and a yellow sari eavesdropping on our conversations with total and unashamed intent. Kar is talking enthusiastically about how the village has been totally cleaned up now, that once you get a toilet, you can't tolerate a dirty environment any more. Drains have been emptied, hand-pumps repaired. Everything is clean and swept and dignified.

I interrupt him to ask the woman why she's so interested. She is pleased to be asked. 'It is good to listen to these things. We are learning how to be clean. There are no flies or mosquitoes now and our children are healthy. When you walk and you don't see shit, you don't get a dirty feeling.' She says that an eviction order has been served on the slum, but that they will be given new, legal land, and that the effects of CLTS will not be lost in the move. 'People will construct tip-top toilets in the new place.' Of that, she is certain.

Shanti Nagar, Mumbai, India – *Author*

9. IN THE CITIES

Aqua-privies for Jesus

At the entrance to Shanti Nagar, off Mumbai's western expressway, you can learn all you need to know about slum-life without having to look hard. In the bustling main street, leading up the hill away from the expressway, men in suits are going about their business, and women are carrying water containers, hurrying to fill them in the two hours that water comes, and the shops and doctors and quacks and barbers are doing good trade. It looks and sounds like any Indian street, though narrower and darker, and you wonder what makes it a slum. So you turn around and see on the dusty roadside, as the cars zoom past, a series of little children's bottoms, perched nakedly and shamelessly in public, defecating with composure, before the children jump up and scamper down the embankment, disappearing back into the slum, and you notice that the roadside is dotted with faeces, for as far down as you can see.

You walk into the slum, and take any of the alleys that run off the main street, and these turn into other alleyways, narrower and murkier, until you feel like you are Alice in a slum wonderland, bewildered and overpowered by the crowds, the noise, the smell, until you feel like you've never been anywhere clean, and you remember how much you love soap.

Cities are supposed to make sanitation easier. They are denser, and their people are generally wealthier than rural inhabitants. It is easier to lay pipe networks because the distances are less daunting, and people generally have enough money to pay for them. Sanitation is a central feature of the concept of the city, at least since the days of the nineteenth century when the medieval, chaotic urban environment was tamed by sanitarians and engineers, and the city came to be defined as a living environment that successfully separates humans from their waste. Urban historians refer to this new urban template as 'the sanitarian city', or, if they're more engineering-minded, 'the hydraulic city'. Even the engineers don't call it the brick or road city, because it was sanitary infrastructure that was the mark of successful urban living. It made successful urban living possible.

Slums defy this logic. They defeat urban planners. They are so shifting, changing and chaotic that experts don't even dare give them a firm definition. Some slums have infernal alleyways; others have sandy streets and banana plantations. Some slums are nice. Some are the worst human settlements imaginable. The compilers of the 1913 edition of Webster's dictionary knew that a slum was 'a foul back street of a city, especially one filled with a poor, dirty, degraded and often vicious population', but the writers of the groundbreaking UN Habitat report *Challenge of Slums* chose instead 'an operational definition'. This attempted to restrict itself to the physical, the quantifiable and the concrete, though actually it defines a slum not by what it is but by what it doesn't have: secure housing tenure, good quality housing and adequate access to water and sanitation. (The report also included one of the better acronyms in urban theory. BANANA: Build Absolutely Nothing Anywhere Near Anyone.)

By some of these definitions, Shanti Nagar is not a slum. Some houses here are worth more per square foot than posh Mumbai apartments. They have two floors and lockable doors. Water is available through a community system: anyone who pays a membership fee gets to use a stand-pipe during the two hours a day water flows through it. The stand-pipes don't need to have taps, because water is either there or it is not, and the most popular containers to store water are old chemical drums which people buy from the recyclers, and which can't ever be safe. But it is water, and it is available.

Shanti Nagar's location is also enviable. It is near a main road, a station, market and schools. It is, for the servants and tradesmen and gardeners who live here, a convenient place with good access to their places of employment. In fact, Shanti Nagar is a desirable place to live, except that not one house has a toilet, inside, outside, or anywhere nearby. The four thousand people who live in Shanti Nagar have to share the facilities of one community toilet block, which has twenty-six stalls. You'll have to ask directions if you want to see one, because smell is no guide when the whole place stinks, and when you can easily be lost down the unlit passageways where you stumble clumsily and where kind hands occasionally reach out and guide your head away from an overhanging tin roof, or move you out of the way of a pile of filth. Shanti Nagar is definitely a slum, because to live in a slum is to live amid shit, and it has plenty of it.

There are office workers and university lecturers living in these slums too. There are houses that gleam with cleanliness. But step outside, and you cannot avoid dirt. It fills the drains, the streets, from the centre to the periphery. But it takes a half hour to walk to the toilet block, a half hour in line and a half hour back, and there are always queues, and you've not been in there five minutes before someone starts knocking. It's easier not to bother. Women who live in slums learn to eat and drink less. Children learn to squat over open drains in the company of flies, crows and vultures. Everyone learns to take a couple of bricks along when going for squatting, for a token elevation over the endemic, awful muck.

This is an extreme version of city living, but it is less and less unusual. We are now an urban species. At some point recently, it is thought, the majority of the world's population became city-dwellers for the first time. A worryingly large part of that city-dwelling in the developing world is in 'unplanned areas', 'informal settlements' or whatever terminology is chosen to describe places like Shanti Nagar. Already, there are nearly a billion slum-dwellers on the planet, and Africa's slum-dwelling population is expected to double on average every fifteen years. In *Planet of Slums*, Mike Davis writes that 55.5 per cent of Indians live in Shanti Nagars, and 85 per cent of Kenya's population growth between 1989 and 1999 happened in its slums. Africa's slum areas are growing twice as fast as anywhere else. One hundred thousand people move to slums every day. The cities of the future will include those made of steel and sophistication, but also living environments constructed from corrugated tin, plastic sheeting, open drains and desperate determination.

Aid agencies have up until now concentrated on rural poverty. Slums are too chaotic, too troublesome. They are not neat. It is no wonder that slums defeat urban sanitarians, who still struggle with how best to dispose of human excreta in the most advanced cities of the world. Countless coastal cities – Vancouver, Brighton – have no better solution for disposing of their excreta that putting it in the sea. What's the hope for slums?

On the main street in Nehru Nagar, another slum built on bad land near Juhu beach, I go looking for Mr Shankar. I'd met him a couple of years before on an assignment. Then and now I was accompanied by the Indian photographer Rajesh Vora, who lives in Mumbai and therefore lives amongst slums, though if he weren't a photographer, he probably wouldn't visit them. Who would?

Shankar had founded Nehru Nagar's Humanist Movement, a local community organization that was secular and unaffiliated with any of the powerful local political parties. He was a good organizer because he was well-known and liked, having lived in this slum all his life, since his father had arrived from some drought-stricken village. When I first met him, he was running a small restaurant where prices were a quarter what they were beyond the gates of the slum. He lived with his family in the alley next to his restaurant. They had one dim room for six people, smaller than the average American parking space, where the only light came from the gleaming tin tableware stacked on a shelf. But the room was as pristine as the alley outside was dirty. The Humanist Movement was a way, Shankar had told me then, of organizing people. It worked, up to a point. The movement's members had marched through Nehru Nagar with a giant mosquito to demand electricity for fans that would keep the mosquitoes away, and electricity was given. They asked for the the *nala*, storm drain around which the slum had grown fifty years earlier, to be cleaned, but that wish wasn't granted. Nor were any toilets.

Two years on, Shankar's restaurant is not to be found, and all the alleys look alike. We keep getting lost. Finally, Rajesh stops outside the building where he thinks the restaurant used to be. A man sits inside a dark room, stripping copper wire, careful amongst tangles. He is a recycler, because slums are the recycling centres of most cities in the developing world. (In Cuba, I saw doorknobs made from telephones and hairdryers converted from Soviet fans and paint cans.) In poor places, nothing is wasted, because waste only comes with wealth, and waste can, in the flourishing informal economy of the slums, create it. The recyclers and 15,000 one-room businesses in Dharavi, a Mumbai slum that is the largest in Asia, create an economic output estimated at $1.4 billion a year.

The recycling man points into the room behind. There are abandoned tables and chairs, the signs of Shankar's restaurant, now closed. Please, sit, the man says, while Shankar is found. The restaurant had closed after the floods two years earlier. Most slums are built on land that is of no use to anyone else. Bad, soggy, sloping ground. Nehru Nagar is inundated often, but these were bad floods made worse by the rumour that preceded them. When they came, it was soon after the tsunami of December 2004, and people stampeded. Twenty-two people were crushed to death in the alleyways by their neighbours' rushing, fearful feet.

'It was terrible,' says Shankar when he arrives, but he also shrugs. It is slum life. He either remembers us or pretends to – plenty of pale faces tramp through Mumbai's slum streets for one reason or another – and insists on taking us on a replay toilet tour. The block we had visited last time is closed now and surrounded by trash and shit. But nearby a new toilet block has been built. It is run by a woman called Rahamath Mathre. She says she is a trained architect, but jobs were scarce and running a toilet block earns her a decent living. Her toilet block is connected to a septic tank which drains directly into the *nala*. Even with Rahamath's services, Nehru Nagar still has only 100 seats for a population of 45,000, figures at some remove from the Mumbai city authorities' target of 1 toilet per 50 people.

Shankar doesn't say why his Humanist Movement hasn't yet marched for toilets, but the SRA might have something to do with it. This is the latest in a series of slum redevelopment initiatives. It was launched ten years ago by Mumbai authorities and pronounced an enormous improvement. As the normal response of city authorities is to bulldoze slums (a tactic beloved of Delhi city officials) or to hope they go away, the SRA was certainly innovative. The SRA enables developers to buy slum land. They can construct apartment blocks on it, as long as some of the blocks are given over to re-housing the slum-dwellers who were living on it. The other blocks can be sold for profit. In Shanti Nagar and Nehru Nagar, colourful boards have sprung up showing visions of SRA life: swanky apartment buildings, green gardens, an SUV parked outside, the modern Indian dream.

The reality, as I discover in the slum of Shambahaji Nagar, is different. These SRA apartments are dark; the corridors have no light, the water supply isn't working and the elevators haven't moved for months. Stagnant pools of water in the hallway and cracks in the walls, though the building is only a couple of years old, are testament to cheap construction. Rajesh is not impressed with SRA projects, and neither are officials of the Maharashtra Anti-Corruption Bureau, who temporarily shut the programme down in 2006. Yet in the apartment of Sumitra Ramchendra Darde, there is no dissatisfaction, though they had a slum house before which was better. It had two storeys, while this has one. 'We had everything but the toilet,' says Mrs Darde. 'That was the selling point.' Even if the SRA architects, seeming not to understand their own people, designed apartments with the toilets opposite the kitchen, an arrangement that most rural Indians – which many slum-dwellers were, originally – would find shockingly unclean. It doesn't matter: Mrs Darde had her toilet room moved, and she even bought a flush Western toilet, copying the ones in the houses she cleans every day. She is beamingly proud of it.

In Paromita Vohra's documentary *Q2P*, a slum mason called Sheikh Razakh is interviewed. He's a liability really, because he builds flush toilets in slums with no sewers that empty into open drains. But that's what people want, he insists. They aspire to 'an apartment where there's a kitchen, bathroom, bedroom. People see that and they want the same for themselves, a bigger house with rooms for everything. They can't have all that so they get the big necessity, a toilet.'

The big necessity is a rarity. It is physically impractical, and would be ruinously expensive, for every slum-dweller to have a flush toilet or latrine. Costly sewer lines would have to be lain by sewer authorities who refuse to do so in slums that are illegally occupied, because that might amount to recognizing property rights. Also, the chances of recouping the costs through tax and rates would be slim. Sewers were installed in parts of Kampala, Uganda, in the early 1930s because engineers calculated that annual running costs would be only £850. In fact, they were £2000, and sewers have turned out to cost five times more than the old nightsoil collection system. Sewers and waterborne sewage treatment may be the default option for planners, in rich and poor cities alike, but in these teeming, dense areas, it is as dreamy a vision as that SUV that will never be bought parked in front of gardens that will never be built.

There are other ways, however, to get sewers into poor urban areas: Brazil has been using 'simplified sewerage' for 20 years now. Its smaller pipes can be laid in shallower trenches, and the main sewer is laid in backyards not on the street (a practice that makes for long, expensive house-to-sewer connections). Costs are slashed. Another simplified sewer system in the huge Karachi slum of Orangi has been a glowing success. Nine out of ten Orangi residents – 95,000 houses – now have flush toilets and sewers. It took years, and required dizzying levels of community organizing: each lane of houses had to agree, and also to pay for the lane sewerage, while the authority generally provided trunk sewers. The Orangi Pilot Project is now a superstar in sanitation circles and has been repeated in forty-two other Karachi slums. In India, simplified sewerage surfaces as 'slum networking', though its spread has so far been limited.

In Mumbai, the next best option in slums is the community toilet block. Historically, these have been installed by the municipality or by whichever politician was seeking slum votes at the time. No provision was made for maintenance. No sewers or water were provided. The National Slum-Dwellers Federation (NSDF), in an excellent publication called *Toilet Talk*, sums up these efforts with withering scorn: 'Each system has its own particular agenda, its own donor constituency and its own attitudes towards slums: government slum improvement systems which pay for expensive contractor-built latrines but no maintenance afterwards; political organizations that do toilets only for particular castes; engineers who do only high-tech ferrocement wonders that poor people are scared to enter; Rotary Clubs which bestow "fully-tiled facilities" but don't provide water connections or follow-up maintenance to go with them; rival charities which do aqua-privies for Jesus, and appropriate-technology types who concoct elaborate biogas latrines with noble dreams of turning faeces into cooking fuel.'

None of them, as you'd guess from the tone, have worked. Pay toilets, meanwhile, charged one rupee per entry. Even assuming you don't have diarrhoea, and you only need to go once a day, this adds up to 150 rupees a month, when the daily wage can be a few rupees and many people earn nothing. There is a joke that poor people are the only ones in cities who can't afford to get diarrhoea. NSDF, under the umbrella of a larger Indian NGO called the Society for the Protection of Area Resource Centres (SPARC), thought there might be a different way.

'Sometimes,' *Toilet Talk* continues, 'grassroots activism involves a great deal of scolding and finger-pointing: "Isn't this awful? Isn't that shameful?" [...] This kind of stuff has limited utility. People in power are more likely to pull back inside their bureaucratic shells like bumped turtles, the minute you start pelting with them "awfuls" and "shamefuls".' SPARC realized that they had something else to pelt the government with, and that was a cheaper, better solution. It would still involve the community toilet, but this time slum-dwellers would be asked to contribute to its construction and maintenance. They are experts in the art of survival, SPARC reasoned, and with more plumbers, joiners and artisans per square foot than probably anywhere else in the city, why not use them? If they help build their own toilets, and the state agrees to supply water and sewer lines, then the usual problems – lack of maintenance, political corruption, hassle – can be avoided. *Toilet Talk* summed up this toilet utopia: 'No middlemen, no contractor's profits, no cream for anybody to skim off. These are 100 per cent fat-free toilets.'

Today, SPARC fat-free toilet blocks serve 400,000 people in eight Indian cities. A SPARC employee takes me to see one in the Kamla Raman Nagar slum. To get to the toilet, you walk past small girls squatting to defecate, step over several open drains, remembering that you once saw water that was blue, not this black-grey mire, and you look for a two-storey building, with gaps in the concrete that serve as windows. Once you see the fish-tank, you're in the right place.

The toilet is run by a community committee headed by Rayeen Abdul Sattar. He greets us at his desk, upstairs on the second floor opposite a small hall that will be used as a computer room when they can afford the computers. Other SPARC toilets use the second storey as a classroom or library or video lounge, because it's clean and people are proud of it. A toilet committee in Cheetah Camp slum used the fees to pay for the only ambulance for miles around, in instalments.

Sattar offers tea and biscuits and tells a tale he's told before. His toilet replaced a municipal one that was neglected and decrepit. It opened on 19 November 2002, on World Toilet Day, though not without troubles. The new toilet offered a membership plan to its users: anyone who wanted to use it could join up and pay thirty rupees a month or one rupee per visit. This didn't go down well. Nobody had paid to use the bathroom before and nobody wanted to start. Sattar spent weeks going from door to door. He had to persuade people for whom thirty rupees was a day's wages. He also dealt with 'anti-social elements', thugs who would 'go from house to house with a book saying people had to pay five rupees if they wanted to "go" [defecate] outside'. If people brought water with them when they came to the old toilet, and they hadn't paid the protection fee, the thugs threw the water away. Sattar doesn't explain how he got rid of them, because by now he has pulled out his ledger and wants to show me the toilet's vital statistics. These have been recorded with the detail that comes with idleness or pride. In and among the 310 households on the plan, there are 280 boys below the age of 10, for example, and 275 girls above it. Sattar believes in attention to detail. He knows people have to be persuaded to care and he does what he can to encourage it. Membership fees pay for a caretaker and three cleaners. They have also paid for the fish-tank at the entrance.

I've never been in a toilet with a fish-tank before. I look at it for a while, probably because it looks so fresh, and because after several days of slum-walking I can feel the dirt going up my nose into my brain. The fish serve a design purpose, because the tank acts as a barrier between male and female entrances. SPARC's design incorporated several innovations, including training women masons to build the toilet blocks and consulting with women on how to improve things. Separate queues were requested, as were doors that swing both ways to make it easier for women carrying heavy water pots or toddlers to enter. Children's toilets are another innovation. They are meant to prevent children being pushed out of the way at busy times, and their open-style stalls and smaller latrine pans soothe children scared of big, dark latrines they can fall into. The children's toilets here conform to both design criteria, but they are empty and look unused. Two twin boys with orange T-shirts and identical smiles have been sticking to me through the tour, and they say they don't like to use the children's latrines, 'because people can see you'. There are better hiding spots outside. The tank is supposed to stop this. 'I put it in to encourage people to come here,' says Sattar. 'So they don't go outside. At least maybe they'll come to look at the fish.'

Not all SPARC toilets are such ringing – and fishing – success stories. There's another in Shamabahaji Nagar. In theory, this toilet should run like clockwork. It has all the parts: a caretaker, a community membership plan. The difference is in the organization. The toilet is managed like a poorly run business. It's five years old but it looks three times that age. There are no doors on the stalls and no water in the taps. There is no fish tank. 'We took away the taps because people were abusing them,' says the toilet's young manager, whose uncle is the big boss. I tell Siddarth Shirur about it. He's a young architect who designed over forty toilets for SPARC in Pune and Mumbai. He has seen pictures of his 'path-breaking' toilets recently and he thought they looked run-down, unkempt. 'I'm beginning to think that open defecation is healthier than those toilets. Is all we're doing creating an enclosure for defecating? It should be more than that.'

SPARC may not be to blame for the failure of some toilet businesses. The organization hands over control to the community after a year-long transition. Then the making or breaking of the toilet is up to the people who use it. This is known as 'self-mobilization'. It's not the whole answer. The development writer Jeremy Seabrook writes that 'it would be foolish to pass from one distortion – that the slums are places of crime, disease and despair – to the opposite: that they can be safely left to look after themselves'. Nonetheless, the genius of the SPARC alliance has been to set up partnerships that bother to include poor people. As *Toilet Talk* concludes plainly, 'the politician is not the man to wash the poor person's bum'.

SPARC's business is booming along with the slums. A new contract will see 200 more toilet blocks built, a total of 2000 new toilet seats. It's impressive, but it's also not much. Siddarth Shirur, who will design some of the new toilets, sounds dispirited. What are a few toilet blocks in the face of the ever-growing slums? 'People keep coming to Mumbai,' he says. 'And they just keep coming.' He watched a slum appear under his windows in three months. 'Four walls, an asbestos roof and the house is ready. The Chief Minister wants to make Mumbai into Shanghai! But how do you stop the slums?'

Dar es Salaam is a pleasant city in the pleasant East African country of Tanzania. As a taxi driver takes me from the airport to my hotel, I realize it's the first normal, stable African country I've visited, because the others have been Liberia (just after a big war), Cote d'Ivoire (just before a little war), Burundi (in the middle of a long war), and South Africa (just, well, odd). In Dar es Salaam, people walk on the streets, unlike in Johannesburg, and wait for buses, unlike in Liberia, which has none. But appearances deceive. Tanzania is one of the most studied countries in Africa. It doesn't have wars, it does have beaches, and it's very poor. This makes it an NGO and academic honey-pot.

I have come here to hang out with a human bundle of energy called Steven Sugden. He works at the London School of Hygiene and Tropical Medicine, but although he's been there for several years, he refuses to call himself an academic. His business card gives his title as On-site Excreta Management Specialist, a title he chose and that he declares is 'catchy'. He came into excreta management after years working on water 'because water's even more bent'. Sugden is an inventor and a do-er. If he wants to find something out, he experiments. Some of the experiments he has told me about include: 'Shitting into a bucket for six months to see how it would decompose'; building a still and peeing into it to try to figure out how and why urine smells; and getting an old schoolmate from Bradford to cobble together a pump that might solve some of the most intractable problems of urban living.

Sugden has come to Tanzania because the London School, via WaterAid, has been commissioned by the World Bank to solve a curious sanitation problem. It is curious because in theory Dar es Salaam doesn't have a sanitation problem. On paper, ninety-six per cent of Tanzanians have a pit latrine or some access to a sanitary facility. But it is a hollow figure. Firstly, having a latrine is a legal requirement in Tanzania, so who, answering a government survey, is going to admit to not having one? Secondly – and this is a problem common to all figures and percentages that relate to sanitation, from Tanzania's ninety-six per cent to those 2.6 billion toiletless – saying someone has a latrine doesn't mean they have safe sanitation. Proper sanitation is a system involving containment, emptying and disposal. It's always more than a latrine.

In a paper accompanying the UN's Human Development Report, David Satterthwaite and Gordon McGranahan noted that 'according to official UN statistics in 2000, the urban population of Kenya and Tanzania appear better served with sanitation than the urban population of Brazil and Mexico. But seventy-five per cent of Brazil's households have toilets in their homes connected to sewers (and many Brazilian cities have virtually one hundred per cent coverage of this); in Kenya and Tanzania [...] a high proportion of urban households classed as having "improved provision" have poor quality pit latrines often shared with many other groups, that present many problems for faecal contamination of users and the wider environment.'

What does this faecal contamination consist of? Flying toilets and DIY emptying. The cheapest latrine in Tanzania, and in every other developing country, is a plastic bag. Kenyans call them helicopter toilets; Tanzanians prefer flying toilets. Whatever the name, the technique is the same: defecate, wrap and throw. Anywhere will do, though roofs are a favourite target and alleys are popular. The plastic bag is a step up from open defecation, because it can be done in private, and it is contained, at least in theory. The next step up is a pit latrine, if it is affordable. Pit latrines in Dar es Salaam cost up to $300 each, because they have sturdy concrete slabs and big three-metre-deep pits, and it requires skill to dig them in Dar's sandy soils. Money can be saved by installing 'temporary toilets' instead, where the pit consists of two oil drums stacked one on top of the other or is lined with tyres. Or by leaving off the roof. The walls in Tanzanian latrines, such as they are – I see hotchpotches of corrugated iron, old doors or rice sacks, held together with hope – only reach to shoulder height. Tanzanians with sharp humour call these structures passport toilets, because they show as much as a passport photograph would and as much as a decent toilet shouldn't.

But the cheapest way to save money is to use the help of the heavens to empty the pit. Wait for the rains, divert the flowing pit contents into the street, hope for the best. Because of this faecal contamination, that deceptive ninety-six per cent figure is hollowed out still more by other numbers. These are the numbers of cholera.

The intestinal bacteria *vibro cholerae* is rightly terrifying. It can kill within twelve hours by causing massive dehydration and constant vomiting and diarrhoea. Though it can be easily treated with oral re-hydration salts, cholera victims who don't have clean water or a twenty-cent sachet of salts will soon have agonising stomach and muscle cramps, they will lose up to thirty per cent of their body weight in hours, then they will turn blue, their eyes will go glassy, and they will die.

Since sewers and sewage treatment vanquished cholera in the industrialized world, it has become mostly a disease of the poor, but not always. After Hurricane Katrina in 2005, US government officials warned that cholera was a possibility because sewage works had been knocked out, and the waters that people were wading in, drinking and drowning in were teeming with faecal contamination. They also warned about typhoid, but it was cholera that made headlines. Perhaps cholera still strikes a chord because there is some lingering memory, in the flushed and plumbed world, of what it used to do and how it used to kill, by the tens of thousands, quickly and pitilessly.

Or because untreated, it will kill half the people it infects, for certain. The World Health Organization calls the absence of cholera 'an indicator of social development'. If decent water and sanitation were present, cholera would not be. Cholera means there is something wrong, no matter what the statistics say. Cholera means the city is not working.

At a clinic in the sub-ward of Azimo in Dar es Salaam, in the month of April, I ask two nursing officials how many cholera cases they have had that year. At first they won't say, and I understand their discomfort, because they could be sacked. After a bad outbreak of cholera in Dar in 2004, six environmental health officers lost their jobs. It was their fault the cholera came, said the authorities. It was not because of imperfect disposal systems for sewage, or the shameful lack of safe pit-emptying services, or the abject failure of Dar's government to make the glaringly logical connection between streets swimming in excrement and the regular appearance of cholera. The nursing officers eventually say they had 3000 cases in one sub-ward in a five-month outbreak the year before. They were getting fifty new cases a week. When I repeat this figure to Sugden, he assumes I mean in all of Dar, and even then that it's shockingly high.

Officially, Dar es Salaam's sewage is treated in a series of waste-stabilization ponds. As sewage treatment systems go, ponds are fine if a city has sufficient land and expertise to maintain them properly. These systems clean sewage by passing it through a series of ponds at a carefully measured velocity, giving bacteria enough time to form and feed on the sewage, cleaning it of solids, and – supposedly – pathogens. There are dozens of pond systems in France and Germany, and the Australian city of Melbourne uses waste-stabilization ponds to clean half of its sewage. But the flow has to be carefully regulated, and so do the ponds.

Sugden takes me to some in a slum area called Buguruni. They are undergoing renovation, but trucks are still discharging into them, including one that is pouring a yellowy-green and foul-smelling liquid into the catchment basin. There is no way that these ponds can dispose of all of Dar's sewage, even if they were working. In fact, they serve only ten per cent of the city population and from the sight of them, they serve them badly. For the seventy per cent majority who have on-site sanitation, which is what pit latrines are, it's every method for itself. In practice, most of Dar's excrement is thrown into gullies and alleys. It is on the streets and in the water, and in the cholera outbreaks that occur every year, steady as the rains.

At a garbage depot in the city ward of Temeke, I meet Mkuu Hanje. He's a senior Environmental Health Officer who has worked for the city for twenty-one years. Now, he's had enough because his work is thankless and impossible. Officially, he is required 'to make sure a community's health is ensured through disease prevention'. When the third biggest killer of children in his city ward is diarrhoea, disease prevention should consist of promptly reporting cholera and ensuring that latrines are safe and safely emptied. In practice, he can do neither because 'when you report cholera, you get into trouble'. Even when he does manage to get someone to court, the fines are laughable, because they date from 1959. His boss is a ward councillor and politician loathe to anger his constituents. 'After five years,' Hanje tells me, 'he must go back to people to get their vote, and my job is touching the day to day life of people. When you take people to court who don't have a latrine, and you've given them notice five times, the councillor thinks you are harassing them. It is a conflict.'

There are ways to get a pit emptied in Dar, but none are cheap and only one is legal. I go to find the vacuum-truck operators who hang out at a junction at the end of Garden Road, their blue trucks coiled with ready hoses but idle. My Tanzanian companion Richard, an earnest and helpful WaterAid intern, advises me to stay in the car to counteract the *mzungu* [white person] effect. He thinks the tanker drivers will be hostile or ask for money. Neither is true. They are courteous and chatty, probably because they're bored stiff. They charge at least 70,000 shillings [US$60] to empty a pit latrine or septic tank and not many people have that kind of money in a city where you can buy bread by the slice. Even if their service were affordable, their trucks can't get down narrow slum alleys to reach the customers. A tanker driver called Charles does most of the talking. He is elegantly dressed with shining brogues, because 'smartness is up to a person's nature', and because his employer refuses to buy him gloves or boots. He tells me that kids call the tanker drivers *nyona mavi*, or shit-suckers. Charles laughs about it but it irks. They're doing a service, and one that the city government can't manage to provide. By law the municipality is obliged to remove nightsoil from residential areas. In practice, this hardly ever happens. How can it, when the department has no tankers of its own? (When the city wants a tanker, it hires one of these, like everyone else.)

Because of this, Hanje's job amounts to firefighting. He must always react to crisis, and never has time to prevent it. It is frustrating. He can't understand why there's no health planning, or why Temeke's municipal health budget in 2004–5 allocated only $3000 to on-site sanitation when 630,000 people rely on it ($3000 works out at less than half a cent per person). 'When there is cholera, and there is a scarcity of drugs, the outcry is very high. But anyone in the street who sees their neighbour disposing of faeces in the street will not complain. There is no outcry about sanitation. If people say, there is no water, or the road is very bad, politicians will react. But when it comes to toilets, no-one complains. If you don't have a toilet, it's up to you.' If you empty it in the street, sowing the seeds of the next cholera outbreak, it's up to you. Hanje sighs, and so does Richard. Having just finished his training to be an EHO, he realizes he's just listened to the story of his next twenty-one years. Hanje wants out. 'If I get a good office that wants to employ me as a cleaner,' he says, 'I'd take the job.' Anything's better than pointless firefighting. 'Maybe sanitation can be changed,' he says with profound weariness. 'But I don't know how.'

The pleasant city of Dar in fact consists mostly of slums. At least seventy per cent of Dar residents live in unplanned areas and urban areas are expanding at nearly six times the country's overall growth rate. Dar's slums are quite nice. They do not, in places, feel like hell on earth. They do not, as Mumbai did, make me angry because no-one should live like that (though I might feel differently during a pit-emptying session). Dar's slums have wide main streets and space. There are some trees, and on the main streets, half-sunken tyres serve as benches. Leading off the main streets are the narrow, familiar alleys of slum living and all the poverty that goes along with them.

In Buguruni, not far from the waste-stabilization ponds, I am trying to talk to some women who are standing behind a wooden stall and selling nothing that I can see. I am using an iPod to record, and lay it on the stall. One of the women reaches for it and therefore so do I, but she's only moving it out of the drips from the straw roof, and I feel prejudiced and terrible, but my prejudice is not unusual. City planners routinely use the excuse that slums are havens of criminality, usually just before they bulldoze them, when actually slum-dwellers tend to be more exploited by criminal gangs than anyone else.

The women don't want to talk because I've asked them how they empty their latrine pit. Richard steps in because he understands what the problem is. He says, 'We're not from the government,' and they laugh with relief and tell me how they would empty their pit if it got full, which is illegally. The most common method used to empty pits, at least by people who don't believe in tipping shit into the streets, is to use a *kutapisha*. The word comes from the Swahili word *tapika* meaning to vomit. A literal translation would be 'one who makes the pit vomit'. Most literature I read about Tanzanian pit-emptiers refers to them more delicately as 'frogmen' – because they dive into pits, presumably – though not one person I meet uses that term even in translation. The *kutapisha* is a mythical figure. Ask a Temeke resident where to find one, and they will respond with the equivalent of seeing a man about a dog. 'Oh, you ask around.' Or, 'You go somewhere really poor and ask anyone. Anyone who is desperate enough will do it.'

Here is how to empty a Tanzanian latrine pit. Take yourself, and a colleague if you can pay one, to the pit in question. It can be during daytime, though night may be better, as your client may be embarrassed by your activity and your activity is illegal. First, take a sledgehammer to the concrete latrine squatting slab that covers the pit. Pour in five kilograms of salt to liquefy the hard layers at the bottom. Leave for one to two days. When you return, bring a can of kerosene and pour it over the contents to mask the smell. Put on boots if you've got them, but you probably won't have gloves, and start digging. It should take a day at least. Do not be alarmed if you find foetuses, and watch out for sharp objects and syringes. When you've finished, decant the muck into another hole and install a new slab. For all this work, you can charge the sum of 95,000 Tanzanian shillings (about $80). Replacing the slab costs extra. It won't be taxed, because you are doing a job unrecognized by, but tolerated by, city authorities. If you didn't do it, no-one else would.

This information is worth money because it comes from a professional. With some effort, Richard has tracked down a full-time *kutapisha* called Mawazo. He's a young man, with smart trousers and confidence, and he talks freely as our little delegation stands in front of his family passport toilet, long enough for the children who gather at any conversation to get bored and move off to find nothing else to do. It's good of Mawazo to take the time, because this is the rainy season when the water table is high, pits fill, and he is most in demand.

Mawazo doesn't use the word *kutapisha*. He says people call him a *kupakuwa*. After some discussion – 'a serving removal man; a person who takes things out of one dish and puts them into another' – this is finally translated as 'pit-scooper or ladler'. The scooper doesn't mind his job because the money is good and he is in demand. Also, though I've heard reports to the contrary, he doesn't have to work at night. Why should he? He has nothing to be ashamed of.

Sugden is impressed. 'You're doing an important job for public health,' he tells him, though Mawazo looks bemused. But it is an imperfect and possibly dangerous solution. By decanting the pit contents into another hole nearby, the *kutapisha* method keeps faecal contamination in the community where it can easily seep into groundwater. Also, Mawazo only empties six pits a month in high season and the contents of other full pits must be going somewhere. Sugden had another idea. If they had a service that was more affordable, that avoided breaking the slab, and that was more accessible for these back alleys, maybe people would use it. Perhaps they would use the Gulper.

Outside the conventional toilet industry, the world of sanitation does not lack innovation. There are thousands of latrine designs; countless varieties of wastewater treatment methods; sewer robots and reverse-osmosis membranes. There are ultra-light wilderness toilets that can be carried in a backpack, and Jiffy bags of crystals for drivers stuck in traffic to pee in. There is even an enterprising retired Navy commander called Virginia Ruth Pinney who has taken the stinkiest compound in faeces – skatole – and weaponized it, according to US patent 6,242,489. Faecal stink bombs are now available as non-lethal weapons for 'riot control, to clear facilities, to deny an area, or as a taggant', according to the arms control group the Sunshine Project. But not many of these innovations had the same research and development process as the Gulper. According to Sugden, this involved, 'oh, about five pints in the pub'.

The Gulper is a manual latrine-emptying pump. In Sugden's vision, it could be carried by one man, who would transport the pump and the emptied pit contents on a simple motorbike. First he needed someone to make it and he knew who to ask. Suggy – his school nickname – was still friends with Stephen Ogden, a farmer in Yorkshire whom Suggy calls Oggy, who gave up dairy farming when it became cheaper to kill a cow than send for the vet, and now does fabricating instead. I later go to meet Oggy at his farm near Bradford. It perches on a patch of rural-ness that takes you by surprise, coming out of nowhere at the top of a normal built-up hill of houses and culs-de-sac. There is a large workshop, some green fields, and a Methodist church that his grandfather built in the farmyard, and which still gets a two-strong congregation on Sundays.

Oggy is all a Yorkshire farmer should be: dry of manner and of wit with a quiet strength to off-set Sugden's energy. Suggy is untethered; Oggy is grounded, enough to say immediately, when I ask if he'd like to go to Tanzania to see the Gulper in action, 'No. No chance. Why would I?' The genesis of the Gulper comes from the fusion of this friendship. It began one sunny April day in 2006 when Oggy was standing by the gates to his farm and Suggy arrived by car. 'He must have talked for two hours straight about shit,' says Ogden. 'I hadn't seen him for a while so after a bit I said, "By the way, how's your wife and children?"'

Sugden explained what he wanted. He showed Ogden some pictures of water pumps to inspire him. They Googled for some more. And after a few months, Oggy had a pump and Suggy got testing. First, they tried it on a barrel of cow muck, but it kept getting clogged. Oggy changed the valve to a hinged one (he tells me this as if it's impossible that I wouldn't understand engineering), and Suggy took it to Tanzania and tried it on some pits. The second pit caused some trouble. 'There was lots of cloth in there,' Sugden tells me, 'and plastic bags from flying toilets. We started pumping and a pair of underpants got stuck in the second valve. We couldn't push it up or down so we had to take it to bits. I looked down at my shoes and my hands covered in shit and thought, oh well, all in the name of research.'

Oggy fitted the pump with an underpants-proof grill, and Gulper 3 was tested, again on cow muck. 'I can still see Suggy teetering above the manure with a stick to stir it and make it thicker.' They put in drier cow manure to account for other pit latrine contents, to give the Gulper a proper challenge. It worked.

The Gulper – so named because 'that's what it sounded like' – may be nothing fancier than a simple stirrup pump with underpants-related features, but it's good enough for several copies to have been ordered by Oxfam. Oggy is bemused by this, but he made the first few pumps. He won't need to do them again, because the point of the Gulper is that it can be copied. I see this attitude often in sanitarians. Dr Pathak of Sulabh; Joe Madiath of Gram Vikas; the two Steves. None have applied for patents. None wants to remove his useful ideas into expensive inaccessibility. This generosity has a fine historical precedent: Dr John Snow, the great Victorian doctor who solved cholera, never patented any of his medical advances despite being a good enough ether practitioner to be requested by Queen Victoria when she gave birth. Patenting is daft, according to the two Steves. It defeats the purpose. 'The idea,' says Sugden, 'is to develop something a small-scale sector can afford and adopt. If you patent it it's expensive and they can't adopt it. It has to be simple and rugged and bomb-proof.' The main technology was ready. Now he had to figure out how to make a business out of it.

Sugden's Gulper project comes under a banner called 'sanitation marketing'. This concept arose from work done at the London School of Hygiene and Tropical Medicine in the early 1990s and derives from the more established discipline of social marketing, which was developed in the 1960s. To an outsider, social marketing might seem like simple common sense, but in the development world, it was a breakthrough. If poor people were treated as consumers, and if things they needed – malaria nets, oral rehydration salts – could be transformed into desirable marketable products, then perhaps poor people could be made to want things that they needed. Sanitation marketing does the same for toilets. Understand how and why people buy latrines, then make it much easier for them to do so. In official Sugden language, this translates into 'to understand the natural acquisition curve and accelerate it'.

One of the foundation stones of sanitation marketing was Dr Marion Jenkins' research. During the early 1990s, Jenkins spent several years asking the people of 520 villages in Benin what made them buy a latrine. She discovered that 'as with any innovation, households will not adopt in a uniform manner and the categories of innovator, early adopter, late adopter and laggard are as relevant to latrine building in developing countries as they are to the adoption of compact disc players or mobile phones in developed countries'. Other studies found that consumer durables introduced into the US before the Second World War took about eighteen years to reach 'take-off,' or proper commercial success. Communications and technology have since brought down this acquisition period to about six years in developed countries. In poorer markets, the timescale remains pre-war. The goal of sanitation marketing is to take that eighteen years period and speed it up using conventional marketing techniques like research, promotion and improved supply chains. In short, in the words of a staffer at the WSSCC, to make toilets like toothpaste. Widely available, cheap enough, and wanted.

In Dar, Sugden's goal was to make the Gulper part of a small business venture. He picked a small enterprise in Temeke called

Tedegro, which already provided decent garbage-removal services. WaterAid would buy the transport and the tank and provide the Gulper. Tedegro had to make the business work.

It sounds straightforward. So had WaterAid's first venture into sanitation marketing in Dar, in the densely populated low-income area of Keko Mwanga B. The idea was to make toilets more available, so why not set up a toilet shop? It would be staffed by a team of *fundis* [masons] and offer a select few toilet designs. It would be accessible, with a location on Keko Mwanga B's main thoroughfare (not really a street, as it turns to mud in the rains). WaterAid followed the principles of sanitation marketing first by doing its homework. A market research study was commissioned that asked the residents of Keko what they wanted in a toilet. They said they were tired of their latrines, which were smelly and unhygienic. They said they wanted something to change because even if they had latrines, their ground-floor houses still suffered when the neighbours emptied their pits illegally. 'What's the point of having a toilet,' they asked reasonably, 'when shit still runs through the kitchen?' They said how sick they were of treading on plastic bags that were helicopter toilets, crash landed on their path. A toilet shop was a great idea. Of course they'd use it.

WaterAid went ahead with the project. A couple of demonstration toilets were built. But when I visit Keko a year on, a toilet *fundi* tells a gloomy tale. People may have had the intention to buy toilets, but intention isn't a sale. 'Most people in the report thought the toilet centre would be providing free toilets. The ones who got demonstration toilets had to pay a little money and so now everyone thinks they can get a toilet like that by contributing an iron sheet or a door.' In fact, as Sugden tells me, the toilets were too expensive. Because the cheapest model was still $270, it offered no advantage over what was already available. Keko's project may have failed but it gave Sugden an idea. He estimated that sixty-two per cent of the cost of the latrine went to digging the pit, lining the hole, and casting the latrine squatting slab. If the cost of the toilet were to be reduced, then the pit had to get smaller too. The Gulper system would not only be able to access narrow slum alleyways, but it would be more affordable because it would offer customers the option of only emptying part of their pits. Sugden calls this 'a pay as you go approach'.

Tedegro's boss, a man called Mohando, is an ex-government official. Consequently, he has good connections that smooth the running of his garbage disposal business. This is run out of a tiny wooden shack on a slum street in Temeke, where the Gulper is temporarily serving as a coat-stand. Mohando has agreed to run the Gulper business in principle, but first he needs the nuts and bolts. Sugden has come to Dar to help source the equipment, and I tag along. Not from any interest in Tanzanian aluminium factories, but because I want to see how a big vision – to remove the seas of shit that drown so many cities – can be built from little things. The complexities of urban planning make my head spin. But I can grasp a tank, a motorbike and a pump.

A *piki piki* – a motorbike with attached trailer – has been sourced in a suburb of Dar. Finding the right tank to attach to the trailer is trickier. Aluminium is the first suggestion, but after several hours of searching, Sugden learns that there are no *fundis* in Dar who know how to weld it. The mood is demoralized and lunch is suggested. Richard proposes a restaurant that is 'conducive'. I don't know what it's conducive for, but probably not for four men to earnestly calculate how much excrement the residents of Temeke produce. Once seated at the table, Sugden gets out his mobile phone and does his sums. One million families, with six people per family on average, producing 200 grams of matter per day on average. That makes 140,000 kilograms. It would take 23 tankers a day to get rid of all that, or 280 trips a day by the *piki piki* to the waste-stabilization ponds. To break even, they need to calculate the number of trips that can be made (five, probably) and a decent salary for the *piki piki* man (5000–8000 shillings a day).

Other considerations must be made. The tank should not be transparent 'because we don't want people to see the shit slopping around'. Sugden has told me the story of the early Gulper experiments, when they emptied the pit contents into dark-coloured plastic barrels and closed them too tightly. Dark attracts heat; excrement produces gas and pressure. The tanks exploded all over Sugden, and the WaterAid driver who had to take him back to the office has yet to get the smell out of his car or to forgive him. With this in mind, Mohando is worried about pressure. Should the tank be totally filled? Will it give? He asks Sugden with the expectation of a man asking an expert. Sugden is an expert, but only so far. 'There's so much we don't know about shit,' he says sometimes. So much to learn. He tells Mohando the truth. 'I don't know. This is new. You'll have to experiment.'

They order lunch and keep calculating. Other pit-emptying arrangements have been tried in Dar and elsewhere. The Vacutug, a gas-engine-powered pump, was devised in 1996 by UN Habitat. But it only travels at walking speeds, says Sugden, 'so it can take ten minutes to empty the pit and two hours to take to the waste-stabilization ponds'. The MAPET was a human-powered pump, but it required three men to operate it, making for punishing running costs (in Mozambique, seventy-three per cent of overheads were salaries). I see a MAPET standing stationary on a Dar street, and hear that its business is bad.

Sugden thinks the Gulper will beat that. It travels at fifty miles an hour, enabling much quicker round-trips. The answer is in the depth of the pit. The *piki piki* could offer more flexible options: two barrels emptied, or four, with scale pricing. 'It needs to be a profitable enough business for them to get another *piki piki*,' says Sugden, 'and to pay for promotion and marketing, but they can't do that until this gets off the ground. It's a chicken and egg situation.'

*

In Dar, I sit on a plastic bench in the yard of Simba Plastics, one of East Africa's biggest plastic manufacturers and suppliers of the ubiquitous rainwater-collecting SimTanks that dot Dar's rooftops. The Gulper gang is hidden behind some water butts, sourcing a plastic tank to fit the trailer. They find one that is perfect and break out in smiles, even the gloomy Tedegro boy who has said nothing for two days. Their enthusiasm is great, but so are the obstacles. Even if the business does work, it can only transport its waste to the already loaded waste-stabilization ponds, where it will probably be dumped in the sea or in landfill. 'You have to make a choice,' says Sugden. 'The argument is, what's worse, contaminating the environment or contaminating human settlement? Obviously the answer is we want to do neither, but it's a very difficult choice.' Pinned to the notice-board at WaterAid's offices in Dar is a cartoon of a man standing by his hut at the base of a cliff, on the top of which a huge boulder is poised to fall on him, while another man says to him, 'Of course your main problem is nitrates in your water.' Priorities. Another cartoon shows two young men standing in front of a sign telling them to prevent cholera by washing their hands before they eat. 'That doesn't apply to us,' says one to the other. 'We don't have any food.' 'No,' agrees his friend, 'and we don't have any water either.'

Sitting on my bench next to some giant Coke bottles – Simba makes soft drink containers too – I consider that solving sanitation is as complicated as it seems simple. It requires the juggling of priorities and budgets. It needs the reform of some local governments and the education of others, like Hanje's short-sighted Ward Officers. I think of how I have got a habit by now of asking all the sanitation professionals I meet how to solve the sanitation problem. I want a magic bullet. Nobody has one. They suggest flexibility is the answer. Adapting solutions to the context. Trial and error, and hoping for fewer errors. I think, can a pump with an underpants-proof grill really do any good? The field of sanitation is littered with pilot projects that die soon after birth. Other manually operated emptying systems – Micravac, Maquineta, Minivac – have never gone beyond trialling stage. At one point during the search, as the rain poured down, Sugden said with no noticeable emphasis, 'We are chasing dreams.'

A few months after I get back from Tanzania, I contact Sugden to ask how the Gulper is progressing. He replies that Tedegro is making a modest profit. He's pleased at how well it's all going. He's got plans to develop an IKEA-style flat-pack latrine. He's been wondering about making fertilizer briquettes out of sludge, and tells me that Oggy thinks a machine he's devised to compress pork belly scrapings might be suitable. He sounds as energetic as usual, but his energy is tarnished with frustration. 'Oggy and Suggy are a sticking plaster team running on a shoestring budget. I'm really struggling to understand why there's not a huge outcry about conditions in [areas like Temeke], why nobody with money is interested or even recognizes it as a problem.' If that continues, he concludes, 'it will always be a sticking plaster approach'. Always one man and his dreams, no matter how grand or how gulping.

Buzz Aldrin on the moon, 1969 – *NASA*

10. THE END

Small energy

The most expensive toilet on earth is designed never to be used on earth. At a cost of $23.4 million, the toilet designed for NASA's space shuttles may seem a ludicrous waste of money. It probably exasperates NASA critics, who see little purpose in spending money on space when there's much to be sorted out on earth. It wouldn't impress Dr Bindeshwar Pathak of Sulabh either, whose handbook complains that 'our scientists think of going to the moon, [but] toilet is not in their vision at all'. Such criticisms are wrong. NASA's attempts to improve the disposal of its crews' excreta in the skies could lead the way for the earth-bound to do the same.

The Environmental Control and Life Support System (ECLSS) that controls the living environment on shuttles and on the International Space Station doesn't have the luxury of disposal, when discharging trash into space has long since been judged a bad idea. In the past, astronauts' conditions were considerably more primitive: when Allan Shepherd set off for the first Mercury shuttle flight on

5 May 1961, no provision was made for any excretion, as the flight was supposed to last fifteen minutes. But it was delayed by four hours, and Shepherd was finally given permission to pee in his space suit after mission control had concluded that he wouldn't damage its precious fabrics.

At that point, says Amanda Young, curator of early space flight at the National Air and Space Museum in Washington, DC, NASA realized 'it was a very real problem'. Faecal bags were developed for the Apollo missions. These stuck to an astronaut's backside, were sealed with Velcro after use then stored until landing. Urine could be dumped overboard, but a hole big enough to dump faeces in space could make the spacecraft too vulnerable. 'If you have a break in the skin of the craft,' says Young, 'oxygen is sucked out of the astronauts. They begin to boil. They'd die in twenty seconds.' For the moon landings, all astronauts were wearing 'faecal containment devices' – like padded shorts – as well as a urine collection bag attached to the suit with a valve. No-one used the faecal options, but a famous photo of Buzz Aldrin is known in certain circles as 'Buzz whizzing'.

At Star City, Russia's equivalent of NASA, a guide called Alexandr tells me that on space-walks cosmonauts wear diapers, yet another example of the superior practicality of Russians in space flight. (This reputation was secured for me by the story – confirmed by Young – that an American company once spent millions on developing a pen that could write upside down in space, while the Russians took pencils.) I tell Young I've heard Russian toilets in space are better and she believes it. 'They build things that are simple and hardwearing.' The Russian International Space Station toilet is also superior, according to astronaut anecdote, though the specialized training required to use toilets in space – a procedure which requires getting a grip on both the receptacle and the sucking, vacuuming devices that come with it – prevented the Americans from using it.

Asking how astronauts go to the bathroom is one of the most

common questions put during NASA or space museum outreach sessions. To cope with the curiosity, for a while the agency posted a video that featured a fully-clothed volunteer showing exactly how it was done: with a mirror, sometimes. Amanda Young is often asked about it. 'Interest from the public is strange. Women don't care. They think, they worked it out and that's that. Men have an almost unhealthy interest. Children are interested in the poop factor.' What everybody should actually be interested in is the drinking pee factor.

Space scientists are faced with a particular difficulty. Water weighs a kilogram a litre. It is heavy and therefore expensive: it costs $40,000 to transport each gallon up to the International Space Station. They don't want to load a shuttle or space station with extra weight, but they need water. So the ECLSS does what anyone would do in straitened circumstances: it turns urine into drinking water. On future Space Station missions, and on the planned 2012 mission to Mars, astronauts will be drinking their own urine, sweat, breath and tears because they have to. When NASA bought a Russian second-hand space station toilet for $19 million, it was not only because the Russians make hardier kit, but because the toilet can recycle urine. Officially, this process is called re-use or reclaiming, and it may be the future of the planet. In fact it's already happening.

Water, water everywhere and none of it infinite. Water is a fixed commodity. At any time in history, the planet contains about 1386 million km^3 of it. Most is salty. Only two per cent is freshwater and two-thirds of that is unavailable for human use, locked in ice, snow and permafrost. We are using the same water that the dinosaurs drank, and this same water has to make ice creams in Pasadena and the morning frost in Paris. It is limited and it is being wasted. In 2000, twice as much water was used throughout the world as in 1960. Water consumption is currently at about 1700 litres per person per day, but that's an average, and most of it is not going down sinks and toilets, but onto fields, into irrigation channels, sprayed around greenhouses. Sixty-nine per cent of our planet's daily water is used for agricultural purposes, and consumption is increasing. By 2050, half of the planet's projected 8.9 billion people will live in countries that are chronically short of water.

Usage is only part of the problem. We are wasting our water mostly by putting waste into it. One cubic metre of wastewater can pollute ten cubic metres of water. Discharging wastewater into oceans turns freshwater into the less useful salty stuff. Desalination is expensive even for the oil-rich desert states of the Gulf that are its biggest fans. (A brief conversation from a wastewater conference with a sewage manager from Abu Dhabi: 'We use sand filters for our tertiary treatment.' 'Oh, I bet you have no shortage of that, do you?' 'Actually the sand is the wrong size. We have to import it.')

The re-use of wastewater effluent is now being proposed in several water-stressed areas of the world. In San Diego, the so-called Toilet to Tap proposal has been rejected by voters several times, and the same happened in Toowoomba, Australia, where rainfall has decreased thirty per cent in the last thirty years. The 'yes' campaign in Toowoomba was led by Councillor Dianne Thorley, who told a TV interviewer that if she had her way, there would be 'advanced water treatment plants bolted onto every sewage plant in Australia'. She was totally convinced that a system using advanced ultra-filtration, reverse osmosis and UV disinfection, or the best cleansing modern science can provide, would ensure adequate safety. It may do – though those organic chemicals from your shower gel will probably still remain – but it can't filter out natural aversion.

Pro-re-use campaigners have two arguments to counter this. To begin with, recycled effluent is already widely used on agriculture. And it's also widely drunk, though unknowingly. Countless human settlements take their drinking water from the same source into which other countless human settlements discharge their raw or treated sewage. Londoners supposedly drink tap water that has gone through seven pairs of kidneys – probably an exaggeration but based in truth, as London takes drinking water from the Thames downstream of towns which discharge their cleaned effluent into the same river. In fact, several American municipalities already do this 'indirect potable re-use'. The Upper Occoquan Sewage Authority's effluent supplies twenty per cent of the inflow into the Occoquan Reservoir, which gives the residents of Fairfax County their drinking water. In droughts, it can supply ninety per cent, and the sewage authority maintains that its highly treated effluent is cleaner than most water sources that end up in the reservoir. Toilets already go to taps.

In 2007, the French bottled-water company Cristaline launched an advertising campaign that featured the slogan 'I don't drink the water I use' next to a toilet, then a bottle of Cristaline, the purer, non-toilet alternative. The French minister of the environment threatened to sue. It was impressive impertinence from a business as profligate with environmentally devastating plastic as bottled water producers are. But impertinence doesn't seem to matter when global sales of mineral water are $25 billion a year. Americans alone last year bought $2.17 billion-worth of only one brand, though this brand – Aquafina, owned by Pepsi-Cola – now admits on its labels that its pure water comes out of an ordinary tap. Despite this revelation, sales were unaffected, leading a spokesperson at the Beverage Marketing Corporation to conclude that 'the consumer just doesn't seem to care about the source'. Anyway, sometimes sewage effluent is cleaner than drinking water: a flusher tells me that 'it's lunacy. We spend a fortune on cleaning effluent to a high standard, then we discharge it into the river, which makes it dirty again.' Other lunacies may include adding orthophosphorous acid to drinking water [to counteract lead leaking in from old pipes], then having to remove phosphorus from sewage effluent.

Re-use works better when it involves camouflage. This technique is used, appropriately for a militarized country, in Israel. At a presentation at a London wastewater conference, a beautiful woman from Israel's Mekorot wastewater treatment utility, who stood out in a room full of grey suits, explained that they fed the effluent into an aquifer and then withdrew it, then used it as potable water. 'It is psychologically very important,' she told the rapt audience, 'for people to know that the water is coming from the aquifer.'

This is a clever way of getting around faecal aversion. Not having wastewater – and not wasting water – would be better still.

Several hours along dusty roads in rural Shaanxi province in China, there is a one-street village called Gan Quan Fang. It's a place of farmers and of houses hiding interior courtyards behind mighty front doors. Gan Quan Fang had been chosen by Plan, an international development agency, as a pilot village to test ecological sanitation. To see the project in place, if not in action, a Plan consultant directs me to the village headman. In one corner of the main room in his large house are stacked several bags of a type of fertilizer that has long been banned in the West because it makes such good explosive. The headman is out, and his wife is busy making small pancakes in a blackened oven that her daughter-in-law feeds constantly with straw, so that her lungs, I bet, are the colour of the oven. The daughter-in-law leaves the straw to wake up their house-guests, three young women currently having their afternoon nap who are working as 'ecological volunteers' for Plan this summer, during a break from their university studies in environmental science.

Two of the young women give their English names as Jessie and Cissie (it is common practice for younger Chinese to take an English name to add to their Chinese one). They are lovely and tenacious, because it's not an easy thing to convince a farmer who has always defecated in his field to install a toilet indoors. 'They think it is alien,' says Jessie. 'It is very strange for them to have a toilet inside the home.' It is even odder to have one that is set up on a tiled platform and involves a ping-pong ball. The Plan eco-san toilet is a urine-diverting twin-pit. The toilet bowl is partitioned; urine goes in one container and solids in another. Drier faeces are easier to compost. A ping-pong ball kept in the urine hole prevents odour, a cheap and locally available innovation that puts it in the development category of 'appropriate technology'. Urine separation would make sense in the world of waterborne treatment, too. Though it only makes up five per cent of the flow, urine contains eighty per cent of the nitrogen and forty-five per cent of the phosphorus that has to be removed at treatment works. Separating it at source would cut down treatment processes and costs. A urine separation toilet also cuts water use by eighty per cent.

The village headman's toilet is right in front of the kitchen. His was the first to be installed because he's the Party Secretary. Jessie and Cissie finish their tea and take me to see some more. Look up, they say. Every house with an eco-san toilet has a vent pipe and the street is dotted with them, like sanitary chimney stacks. The volunteers won't take me to the houses that haven't installed toilets yet, because they don't want to embarrass the farmers. Instead, we go to meet the teachers. Wu Zhao Xian and Zhang Min Shu both teach primary school. He is tall and thin; she is shorter, wider. Mrs Wu has a wonderful giggle. They serve delicious tea in the kitchen, which is where Plan wanted to put their toilet. Impossible! That was too near food. The toilet is in the courtyard now, a decent distance away. 'We thought it would stink. Now we know it doesn't, we'd have it anywhere in the house.' I ask to use the bathroom, because I'm on my fourth cup of tea in an hour, and the small room, with a curtain for a door ('only for some privacy when we shower') is clean and decent.

Mr Zhang explains how his 'techno toilet' works. 'It's very scientific. There are two solid waste containers. We only need to clean it once a year. Once it's full, we swap the containers.' The contents of the full container are removed, hopefully by now safely composted and pathogen-free, and applied to fields. The empty container moves into the full one's place, and another year should go happily by. Done properly, eco-sanitation turns waste into safe, sowable goodness. Done properly, there's little argument against it. It is sustainable. It closes the nutrient loop, which sewers and wastewater treatment plants have torn open by throwing everything into rivers and the sea, damaging water and depriving land of fertilizer. The arguments in favour of ecological sanitation are numerous and persuasive.

Yet it provokes hostility. I hear references to the 'eco-mafia', to those 'damned Germans and Swedes', the two leading eco-san nations. One sanitation professional says, 'What I hate about this sector is how things develop into fashions and all of a sudden, the answer's eco-san, now what's your problem? But all eco-san wants to do is close the loop. They're not thinking about changing hygiene behaviour.' Pete Kolsky, who admittedly works for the World Bank, an institution that continues to invest in waterborne sewerage for developing countries, is yet to be convinced by eco-sanitation. 'I will give the argument its intellectual due,' he says, 'that it's about nutrient recycling for growing food. The trouble is that people then forget about public health, and if I can't convince a household to invest $50 in a basic cement slab and a pit, what on earth makes you think I'm going to convince them to invest $300 in an eco-san latrine?'

I phone the Norwegian mafia. Petter Jenssen is an agricultural professor at the Norway University of Agriculture and a confirmed proponent of eco-san. I ask him why the debate is so acrimonious. Why do eco-san fans annoy everyone who isn't one? 'The way people present eco-san is often a bit religious,' he says, meaning the fundamentalist kind. 'It's eco-san or nothing but. That can trigger people's resentment. Also early systems did have drawbacks and they didn't see them.' Eco-san if done wrong can leave pathogens in the composted or dehydrated excreta. Even when done well, it may not get rid of worm eggs. Also, it can require huge behaviour changes that are notoriously difficult to achieve. Urine diversion toilets, for a start, require men to urinate sitting down, a shock to anyone used to the ease of what Germans call stand-peeing. Not every man, I suspect, would be as amenable as Mr Zhang in Gan Quan Fang, who is serene about such things. 'For me,' he tells me with a big, satisfied smile, 'whatever the toilet is, I use it. For example, here we eat wheat. When we go to the south of China, we eat rice. Otherwise we starve.'

Jenssen says the technology is better now, and so are the persuasive techniques that can make behaviour change possible. Eco-san advocates have learned that talking about 'sustainable sanitation' raises fewer hackles.

The new refrain in sanitation is 'flexibility'. Darren Saywell of the International Water Association wants to get rid of the 'sanitation ladder', the concept of a sanitary progression that starts with a pit latrine and always ends with the desired ideal of a flush toilet and waterborne sewerage. He thinks the ladder is too linear. The flush toilet doesn't have to be the holy grail of hygiene. The Canadian academic Gregory Rose points to the example of cellphone technology. Developing countries without phone systems didn't bother with telephone poles and underground cables. They vaulted directly to cellphones and satellite communications. Similarly in sanitation, Rose writes, 'the opportunity I see for developing countries is to leapfrog over the dinosaur technologies we have funded and implemented in the North and move to these advanced technologies,' such as composting latrines or waste-stabilization ponds. It is time for appropriate sanitation technology, not blind faith in flushing.

The other fashionable concept is sustainability. This has penetrated even the rich world of engineering certainties and infrastructurally invasive sewers and wastewater treatment plants. At the wastewater treatment conference I attended in London, the biggest draw by far was a presentation by Duncan Mara about waste-stabilization ponds. Afterwards, I asked Mara why he was so popular. He wasn't certain, but thought that it was probably due to panic. Wastewater treatment plants now have to conform to strict EU energy limits, and ponds don't require electricity. Government has finally noticed that wastewater treatment is not green. A sewage works uses up 11.5 watts of energy per head, or a quarter of the output of the UK's largest coal-fired power station. The UK's environment minister made this point in a new national water strategy, when he wrote that 'there's a carbon impact here that simply has to be tackled'.

Other things will also have to be tackled. Pharmaceuticals in wastewater will be the next headache. A recent investigation discovered that the city of Philadelphia utility found ninety per cent of the drugs it tested for, including evidence of medicines used for heart disease, mental illness, epilepsy and asthma. These are tiny, trace elements, admittedly. Studies have shown that pharmaceutical residues have made male frogs produce eggs and deformed some fish. As for the effects on humans, no-one yet knows. But being exposed to other people's excreted drugs in drinking water, said one environmental health professor, 'can't be good'. A senior EPA official admitted that 'there needs to be more searching, more analysis.'

David Stuckey, an engineering professor at London's Imperial College, was recently awarded nearly a million dollars to develop an innovative anaerobic process that reduces sludge volumes by ninety per cent. This will save enormous amounts of money and considerable carbon dioxide. He thinks change must come, and it will be through economics. 'People are looking to invest in wastewater treatment,' he tells me. 'You don't have to be a genius – just look at the price of resources and the cost of nitrogen and phosphorus. Once costs go up, people change.'

Petter Jenssen sits on the other side of the waterborne

sewerage/ecological sanitation divide, but he agrees. 'We've invested so much in conventional sewerage. There are many economic interests tangled up in it. It depends on what politicians dare to do. Maybe it will take another fifty years to reach a sustainable system. But things can happen. Fifteen years ago I was considered a romantic scientist. Now I'm chairman of the national Water Association.'

In the toiletless world, there is cautious hope. The Prince of Orange, chair of the UN Secretary General's sanitation advisory panel, is a significant man, says Eddy Perez of the World Bank, and he represents an increasingly significant cause. 'Not only is he royalty,' says Perez, over coffee in World Bank's gleaming lobby, 'but he knows what he's talking about. He comes to the Bank and has meetings with vice-presidents, to say, what are you going to do about sanitation? Two years ago that would have been completely unbelievable.' An outsider might think that the prince gets such access because he's the figurehead of the International Year of Sanitation 2008. This must be a big deal. In fact, the International Year of Sanitation nearly lost out to the International Year of the Potato when potato-growing New Zealanders objected. Now there are two dirt-related years on offer. Toilets or tubers.

UN years are notoriously meaningless. This one may seem remarkably empty when the sanitation Millennium Development Goal is currently the furthest off-target. To meet its goal, 95,000 toilets must be installed every day. One toilet per second, every twenty-four hours until 2015. Nonetheless, Pete Kolsky, who describes himself as a glass half-empty man, thinks something has shifted. 'This sector has finally got its act together. I don't think it's widespread among politicians yet but it's getting there. They're getting a clear and coherent message that goddammit, it's time to talk about crap.'

People who work in sanitation sometimes have visions. Eco-san people see a future where instead of controlling pollution after it happens, we prevent it in the first place by some kind of source separation. Water separated from excreta; urine separated from faeces. The discarded products of the human body given treatment appropriate to one name (shit, meaning to separate), not another (waste, from the Latin *vastus*, meaning unoccupied or uncultivated). I read about a cleaner new world where people put out their bins full of faecal compost to be collected on a Monday, like they do with garbage.

Why not? Recycling is relatively new yet pervasive (though not necessarily much use and rarely lucrative). Three thousand Swedes have NoMix urine separation toilets. Large-scale eco-san projects in Linz, Austria and Dongsheng, Inner Mongolia are working, more or less, though the Austrians are hampered by legislation forbidding them to re-use urine, and the Inner Mongolians have a bad habit of cleaning their urine diversion toilets by flushing water down them, defeating their source-separating purpose. These visions are seductive from an environmental perspective, but I wonder if they are realistic. The flush toilet is ubiquitous because it is useful. It removes faeces from the living environment; it doesn't take up space; it can be situated inside a house; it can be used in multi-storey housing and it isn't smelly. It seems unassailable as the default option of how to dispose of human excreta in sophisticated, wealthy places.

But it can evolve. In a small French village near the house where I wrote this book, there is a tiny public toilet on a bridge. It has been replaced by newer toilets that do not, as this one does, dispense with sewage by means of a big hole and the river below, and is locked and unused. But on the wall outside the toilet, someone has fixed copies of the original designs and documents that were drawn up to build the toilet a hundred years earlier. The documents are a sign of pride, and rightly. I like engineers. They build things that are useful and sometimes beautiful – a brick sewer, a suspension bridge – and take little credit. They do not wear black and designer glasses like architects. They do not crow. Many of the world's waterborne sewage systems are engineering marvels. The public health benefits brought by modern wastewater treatment are undeniably stunning and still persisting two centuries on. There are plenty of reasons to support waterborne sewerage, to be a fan of flushing. On the riverside path running alongside Crossness, there is a poster celebrating the work of Joseph Bazalgette. The title reads, 'Great Stink – Great Solution'. But it's the assumption that the sewer is the solution that has prevented evolution.

Since I started researching this book, I've noticed I do some things differently. I always put the toilet seat lid down before I flush because I've learned that urine sprays fine mists. I use less toilet paper and more soap and water. I wash my hands more, and mostly follow the five-step hand-washing guidelines provided by the Centers for Disease Control, which at some point I plan to pin up in most pub urinals and the student bathrooms at the London School of Hygiene. Because I wash my hands more, I use more hand-cream and when I wash my hands again, I think of organic chemicals and pharmaceutical residues going down the drain. Yet I still use the hand-cream and the shampoo and the facial wash and all the other products helping to top up my body burden.

What else? I notice manholes. I scowl at paved-over front gardens. I pour used cooking oil on a flowerbed, in memory of those fat-blocked sewers. If it's not urgent, I don't flush the toilet immediately and I don't feel bad about it. This is what hygiene specialists would call behaviour change. The sociologist Harvey Molotch prefers to call it 'the moral rectitude of pee'. He explains: during a five-year drought in California, people stopped flushing urinals to save water. 'I was in the bathroom one day with a colleague from overseas who said, "That's disgusting". It hit me that it's not disgusting. In a drought context, darker yellow pee in urinals signified moral rectitude.' Petter Jenssen tells me about the children of a friend. 'He had put in a composting toilet and when his children started kindergarten they didn't want to go to the toilet because they could see shit floating in the water and they were used to a black hole. You can get used to anything.'

It's no use dismissing sanitation as a problem of the poor. It's true that hardly anyone in the developed world now dies from dirty water. But the diseases of the poor now travel thanks to international airlines. In the World Health Report 2007, WHO director Dr Margaret Chan wrote of a disease situation that was 'anything but stable'. New diseases are increasing at the historically unprecedented rate of one per year. Airlines, Chan wrote, 'now carry more than two billion passengers annually, vastly increasing opportunities for the rapid international spread of infectious agents and their vectors'.

To understand what this has to do with modern plumbing, go to Amoy Gardens, a large apartment block in Hong Kong. In 2003, a seventh-floor bathroom in Amoy Gardens became an epicentre of an outbreak of SARS (Severe Acute Respiratory Syndrome), when a man who had caught the infection in hospital came to stay with his brother. He had diarrhoea, and because his brother's bathroom wasn't working properly – the air vent was sucking out sewage droplets trapped in faulty U-bends – his infectious diarrhoeal droplets were spread up and down the floors. Because SARS spread from person to person, required no vector and incubated unnoticed for a week, it spread on several international flights between mainland China, Singapore and Canada. By the end of the four-month outbreak, 8422 cases had been reported and eleven per cent of those died. And some of it happened because of faulty sanitation in the developed, sophisticated environs of Hong Kong. No country is protected, the WHO report stated bluntly, 'by virtue of its wealth or its high levels of education, standards of living and health care or equipment and personnel at border crossings'. In fact, SARS was 'a disease of prosperous urban centres'. And of those centres' apparently sophisticated but actually flawed and disregarded sanitation infrastructure. The slums of Mumbai and Lagos may seem far removed from modern cities built on sturdy sewers and sanitary confidence. They look like a seething, desperate Dickensian past. But what if they're the future?

The economist William Easterly, in *White Man's Burden*, recently divided the aid world into Planners and Searchers. Planners are the top-down bureaucrats; they are the ones who put toilets on people's heads while not thinking of making the nearby bush seem an unattractive option instead. They are the engineers and wastewater managers, the workshop attendees and the grey suits, who don't want to do differently. They are the stagnant status quo of sanitation. This book has been about Searchers. Joe Madiath, Kamal Kar, Wang Ming Ying. Jessie, the Plan volunteer in Gan Quan Fang, who took her leave near a field of apple trees fertilized with biogas slurry by saying, 'I just want to change my surroundings. I know I am only small energy. But I want to try my best.'

What is the best? Someone once asked me for a systems theory about sanitation. He wanted something that could solve everything. I had to disappoint him. But I think the first step is an easy one. In 1940, Harold Farnsworth Gray concluded that 'urban man today still unnecessarily pollutes streams, bathing beaches, bays and estuaries, without benefit of the excuse of ignorance which was available to his ancestors'. We have less ignorance and less excuse. The first thing sanitation needs is a spotlight shining on it. It needs to be unshackled from shame. It needs some scrutiny.

We can do better, I'm certain, than the young man in London, who came around the hours of midnight to where I was standing with the flushers around an open manhole. The pubs had just closed, but he was the only person among several passing crowds to amble over to what I thought was an arresting scene of several burly men and one woman in crotch-harnesses, paper overalls and thigh-high waders, standing around a black hole in the street. The young man asked what we were doing. It was a good start. I waited for him to ask more, such as where the sewers went and what happened when they got there. Or why we refuse to notice that we still don't know how to properly deal with something that we all produce, up to several times a day, many million years after we first started producing it.

But a flusher answered his only question, quick as a rat, with, 'We're opening a nightclub, mate.' And the young man nodded and moved away, with no further questions.

Acknowledgements

Thank you, firstly, to everyone who shared their stories and showed me their latrines with no embarrassment and impressive pride. Thank you to my interpreters and translators Red Chan, Richard Kishere, Hassan Zeini, Sayuka Tsujita and Motoko Brimmicombe Wood, for their serene professionalism no matter how unpalatable the subject. For their hospitality and help, sometimes in the most inhospitable surroundings (sewers, slums), I am grateful to Kevin Buckley, Toshio Iwama and Chizuku, Joe Madiath, David Miller, Trevor, Audrey, Anza, Masana and Wanga Mulaudzi, Rob Smith, Simon Winchester and Rajesh Vora. To everyone who pointed me in the direction of research material, sent comedy cards, and shared their sanitation history, thank you. Jack Sim, Rob Smith, Caroline Snyder, Steven Sugden, Clara Greed, Douglas Greeley and Pete Kolsky answered endless follow-up questions with haste and good grace. Adam Broomberg, Delphine de Lardemelle, Hamlet, Dennis Delaney and Karen Robinson assisted with images, while Julie Bindel, Melanie

McFadyean, Lisa Margonelli and Mary Roach supplied sterling sideline support. Claire and Julie, emergency nurses at Homerton A&E, provided care, cheer and typing-friendly bandages at a critical time. Thank you.

For their insight and corrective care while reading some or all of the book in draft form, I thank Scott Chapman, Clara Greed, Les Jones, Pete Kolsky, Molly Mackey, Trevor Mulaudzi, Steven Sugden and Tom Ridgway. Thomas Jones at the *London Review of Books* and June Thomas at *Slate* kindly published sewer stories along the way. My agents Erin Malone at William Morris and Peter Straus at RCW were indefatigable, and thank you, Erin, for the Playmobile toilet cubicle, an effective procrastination deterrent. I am indebted to Riva Hocherman, Steve Hubbell and Philip Gwyn Jones for their patient and careful editing, and for their wholehearted and comforting championing of a book about a difficult topic. Thanks to all in Rivel, especially Cathy and Shaun Faulkner, for providing a quiet writing refuge, and to Molly and Maggy for the walks. Thank you, finally, to my family, Sheila, John, Allison and Nicholas Wainwright and Simon George, for being there.

Notes

Introduction

2 *2.6 billion don't have sanitation* – United Nations Development Programme, Human Development Report 2006, 'Beyond Scarcity: Power, poverty and the global water crisis', New York: Human Rights Development Office, Palgrave Macmillan, 2006, p.v.

– *10 million viruses, 1 million bacteria* – Water Supply and Sanitation Collaborative Council (WSSCC), 'A guide to investigating one of the biggest scandals of the last fifty years', undated, p.3.

– *One in ten of the world's illnesses* – Bad sanitation is linked to, for example, respiratory infections (which can be spread by unwashed hands); malnutrition, malaria and worm infections. A. Prüss-Üstün et al, 'Safer water, better health: costs, benefits and sustainability of interventions to protect and promote health', Geneva: WHO Press, 2008, p.8.

– *Nearly ninety per cent* – World Health Organization, 'Water, Sanitation and Hygiene Links to Health, Facts and Figures, updated November 2004,' available from http://www.who.int/water_sanitation_health/publications/facts2004/en/

– *Ten per cent of cells in our body are actually human* – Gary Hamilton, 'Insider Trading', *New Scientist*, 26 June 1999.

3 *Since the Second World War* – 'Responding to the Sanitation Scandal: WSSCC launches the global sanitation fund', WSSCC, news release, 7 March 2008.

– *The biggest medical milestone* – 'Medical Milestones', *British Medical Journal*, http://www.bmj.com/cgi/content/full/334/suppl_1/DC3

– *One child out of two* – Stephen Halliday, *The Great Stink of London: Sir Joseph Bazalgette and the cleansing of the Victorian capital*, Stroud: Sutton Publishing, 1999, reprinted 2007, p.vii.

– *Mortality dropped by a fifth* – Johan Mackenbach, 'Sanitation: pragmatism works', *British Medical Journal* (2007; 334, supplement 2), 6 January 2007, p.s17.

– *Reduce diarrhoea by nearly forty per cent* – Val Curtis and Sandy Cairncross, 'Water, sanitation and hygiene at Kyoto', *British Medical Journal* (2003; 327), 5 July 2003, pp.3–4.

– *Modern sanitation has added twenty years* – Personal communication with Gary Ruvkun, March 2003.

4 *50,000 people nationwide* – Steven Johnson, *The Ghost Map: A street, an epidemic and the two men who battled to save Victorian London*, London: Allen Lane, 2006, p.70.

5 *Forced to issue boil-water notices* – 'We'll be boiling water until the summer – HSE', *Galway Advertiser*, 29 March 2007.

– *600 times levels permitted* – RTE, 'Dirty Water: Galway', 24 May 2007, http://www.rte.ie/radio1/investigate/1134569.html

– *We haven't had any for two years* – Clare County Council, Ennis Public Water Supply Reminder, 14 December 2007, http://www.clarecoco.ie/news/Ennis_Public_Water_Supply.html

6 *Nearly half the country* – Working data table provided by Peter Kristensen, European Environment Agency, based on data from Eurostat and national sources.

– *Fined $15 million a day* – Personal communication with European Commission press office, February 2008.

– *Milwaukee's drinking water* – Natural Resources Defense Council (NRDC) 'Swimming in Sewage', February 2004, p.50; see also Robert D. Morris, *The Blue Death: Disease, Disaster and the Water We Drink*, New York: Harper Collins, 2007.

– *'Full-strength, untreated sewage'* – NRDC, op. cit., p.51.

– *Ninety per cent of the world's sewage* – 'World Sanitation Goals Slip, Nature can Help,' Reuters, 16 March 2008.

8 *'Defecation is very lowly'* – Bindeshwar Pathak, 'History of Toilets', paper presented at International Symposium on Public Toilets, Hong Kong, 25–27 May 1995.

– *'The lonely bewilderment of bodily functions'* – William Cummings, 'Squat Toilets and Cultural Commensurability: Two Texts, Plus Three Photographs I Forgot to Take', *Journal of Mundane Behavior* 1:3, October 2000, p.268.

– *'A great and glorious thing'* – *Letters of Rudyard Kipling* (edited by Thomas Pinney), Basingstoke: Macmillan, 1990–2004, vol. 1, p.121.

– *'A meritorious undertaking'* – Sigmund Freud, 'Preface to Bourke's *Scatologic Rites of All Nations*', in *Standard Edition of the Complete Psychological Works of Sigmund Freud* (translated from the German by James Strachey), London: The Hogarth Press, 1995, p.337.

9 *Slowly through a pond filled with reeds* – http://www.princeofwales.gov.uk/personalprofiles/residences/highgrove/

– *The only patent-holding monarch in the world* – The patent for the Chaipattana Aerator was granted to King Bhumibol on 2 February 1993. The Chaipattana Foundation, http://www.chaipat.or.th/chaipat_old/journal/aug03/aerator_e.html

– *My baby doesn't smell* – Trevor I. Case, Betty M. Repacholi and Richard J. Stevenson, 'My baby doesn't smell as bad as yours: The plasticity of disgust', *Evolution and Human Behavior* 27 (2006), pp.357–365.

– *The Scatological Rites of Burglars* – Albert B. Friedman, 'The Scatological Rites of Burglars', *Western Folklore*, 27 (1968), pp.171–179.

– *Sniffed like snuff* – Martin Monestier, *Histoire et Bizarreries Sociales des Excréments des Origines à Nos Jours*, Paris: Le Cherche-Midi, 1997, p.93.

10 *Ninety per cent of patients given faecal transfusions recover* – Thomas Borody, a pioneer of the technique, has written that 'there are few medical therapies that reduce illness so dramatically'. Thomas J. Borody, 'Flora Power: Fecal Bacteria Cure Chronic *C. difficile* Bacteria', *American Journal of Gastroenterology*, August 1995, vol. 11, pp.3028–9.

– *'You don't ever see or smell a thing'* – Megan Levy, 'Grandma saved by daughter's poo', *Daily Telegraph*, 29 November 2007.

– *'Once people got talking about bathrooms'* – 'Examining the Unmentionables', *Time*, May 20, 1966.

12 *A dozen categories of euphemism* – Steven Pinker, *The Stuff of Thought: Language*

as a Window into Human Nature, London: Allen Lane, 2007, p.351.

13 *Without talking frankly about shit* – WSSCC, *Listening*, Geneva: WSSCC, 2004, p.44.

14 *Humanity's 'wiser course'* – Freud quotes from the last scene of *Faust*, where the 'more perfected angels' lament *Uns bleibt ein Erdenrest/zu tragen peinlich/und wär' er von Asbest/er ist nicht reinlich* ('We still have a trace of the Earth, which is distressing to bear; and though it were of asbestos, it is not cleanly.') Men, Freud writes, 'have chosen to evade the predicament by so far as possible denying the very existence of this inconvenient "trace of the Earth", by concealing it from one another, and by withholding from it the attention and care which it might claim as an integrating component of their essential being.' Freud, op. cit., pp.335–6.

Chapter One: In the Sewers

18 *Sewer workers . . . Joe Harlow and Dave Yasis* – 'Body of Missing Minnesota Sewer Worker Found', *Associated Press*, 28 July 2007.

– *Stalwart good-looking specimens* – Henry Mayhew, *London Labour and the London Poor*, 1851; London, Cass, 1861–1862, vol.2, p.428.

19 *300 flushers* – personal communication with Douglas Greeley, New York City Department of Environmental Protection, September 2007.

– *284 égoutiers* – Section de l'Assainissement de Paris, http://www.paris.fr

– *At the time of my visit, it was thirty-nine* – Thames Water press office says there are now forty-nine full-time sewer inspectors, of whom forty-three are flushers.

– *Its third owner in nineteen years* – 'RWE sells Thames Water Holdings plc to Kemble Water Limited for GBP 4.8 billion (€7.2 billion)', Thames Water, news release, 17 October 2006.

21 *915 million litres of clean water* – 'Mayor sets out case against a desalination plant for London', Greater London Authority, news release, 26 July 2006.

23 *850,000 handsets a year are inadvertently flushed* – 'Brits flush away mobiles worth £342 million', Simplyswitch, news release, 4 June 2007.

– *A bra and knicker set* – 'Flushed bra causes sewer collapse', BBC News, 19 June 2007.

24 *'That's Canal No. 5, that is'* – Sukhdev Sandhu, *Night Haunts*, London: Verso, 2007, p.65.

25 *The contents of his neighbour's privy* – 'Going down to my cellar...I put my feet

into a great heap of turds, by which I find that Mr Turners house of office is full and comes into my cellar, which doth trouble me; but I will have it helped.' *Diary of Samuel Pepys*, 20 October 1660, edited by Robert Latham and William Matthews, London: G. Bell & Sons Ltd., 1970, p.269.

26 *The most unfashionable muck* – The enterprising sewer workers of Paris also skimmed corks from the flow and sold them to perfume makers to use as stoppers. Donald Reid, *Paris Sewers and Sewermen: Realities and Representations*, Cambridge, MA: Harvard University Press, 1991, p.53.

– *'Courtyards are filled with urine and faecal matter'* – *'Le réceptacle de toutes les horreurs de l'humanité (...) les passages de communication, les cours, les corridors sont remplis d'urine et de matières fécales.'* Turneau de la Morandière quoted in Monestier, op. cit., p.98 (my translation).

– *Excreta would corrode the gold* – Dr Aleksandr Lipkov, lecture at the World Toilet Summit, Moscow, September 2006.

– *London grew* – Halliday, op. cit., p.45.

– *The cesspool emptying fee* – Johnson, op. cit., p.10.

– *Tudor environmental health inspectors* – The sewer commissioners were commanded, among other things, 'to reform, repair, and amend the said walls, ditches, banks, gutters, sewers, gotes, calcies, bridgess, streams and other the premises, in all places needful.' 'The Bill of Sewers with a new proviso, etc.,' 1531, in Danby Pickering, *The Statutes at Large from the first year of Richard III to the 31st year of K. Henry VIII, inclusive*, Cambridge: Joseph Bentham, 1763, p.223.

27 *'Sewer' either derives from 'seaward'* – Mary Gayman, 'A Glimpse into London's Early Sewers', *Cleaner*, March 2006.

– *The 3700-year-old palace of King Minos* – International Water Association (IWA), 'Sanitation 21: Simple Solutions for Complex Sanitation', London: IWA, 2006, p.12.

– *'The everyday sanitary conveniences of Minoan Crete'* – Virginia Smith, *Clean*, Oxford: Oxford University Press 2007, p.6.

– *A large city sewer that was cleaned by prisoners of war* – Caroline Schönning, 'Urine diversion – hygienic risks and microbial guidelines for re-use. Background paper for WHO Guidelines for the safe use of wastewater, greywater and excreta', Geneva: World Health Organization, 2006, p.30.

– *'The Thames is now made a great cesspool'* – House of Commons, *Report from the Select Committee on Improvement of the Health of Towns, together with the minutes of evidence, appendix, and index. Effect of interment of bodies in towns. Ordered to be printed, 14 June 1842*, Shannon: Irish University Press, 1970, q.3452, pp.209–210.

28 *14,000 died in London* – Halliday, op. cit., p.124.

– *'What health is'* – Henry Mayhew, letter to the *Morning Chronicle*, 24 September 1849.

29 *In modern money* – The economists Lawrence H. Officer and Samuel H. Williamson, who run the website http://www.measuringworth.com, from which these calculations are taken, set out various criteria for translating expenditure from one period to another. The lower figure of £3 billion comes from a calculation made using the retail price index; the higher figure from GDP. The modern worth dates to 2006.

31 *£6 million a year to remove [fat]* – Personal communication with Rob Smith, February 2006.

– *Many . . . sewers will be 250 years old* – Kirsty Scott, 'The dirty bomb beneath our feet', *Guardian*, 23 September 2003.

– *6000 homeowners* – House of Commons Committee of Public Accounts, 'Out of Sight – not out of mind: Ofwat and the public sewer network in England and Wales', Thirtieth Report of Session 2003–4, London: The Stationery Office Ltd., 2004, p.3.

– *Unintended sewage pond feature* – Sonia Young was awarded £6000 in damages from South West Water to compensate her for seven years of cleaning up 'the toxic, smelly and very sad stream'. 'Compensation award for seven years "sewage hell"', *Cornishman*, 7 February 2008.

– *Sewage that reached her knees* – Laura Matless, 'Mother and baby flee from foul flood waters', *Bath Chronicle*, 14 October 2005.

32 *Sewage backed up . . . once a year* – Opinions of the Lords of Appeal for Judgement in the cause Marcic (Respondent) v. Thames Water Utilities (Appellants), [2003] UKHL 66, http://www.publications.parliament.uk/pa/ld200304/ldjudgmt/1jd031204/marcic-1.htm

– *'Sewerage undertakers'* – House of Commons Committee of Public Accounts, op. cit., Ev. 2.

– *Bonuses totalling £1.26 million* – Martin Horwood MP, *Parliamentary Debates*, Westminster Hall, 27 June 2006.

– *'One of the most unpleasant events'* – House of Commons Committee of Public Accounts, op. cit., Ev. 2.

35 *D minus* – American Society for Civil Engineers, *Report Card for America's Infrastructure 2005*, http://www.asce.org/reportcard

– *Crumbling, dangerous sewer pipes* – NRDC, op. cit., p.23.

36 *2175 Olympic-sized swimming pools* – David Hsu, 'Sustainable New York City', New York: Design Trust for Public Space & New York City Office for Environmental Co-ordination, 2006, p.21.

– *1.46 trillion gallons* – US EPA, 'Implementation and Enforcement of the Combined Sewer Overflow Control Policy, Report to Congress', U.S. EPA 833-R-01-003, December 2001, pp.7–15.

– *1.75 inches of rain falling in an hour* – personal communication with Douglas Greeley, September 2007.

– *3.5 inches of rain* – Ibid.

– *'A design issue'* – 'Intense rain storm cripples NYC mass transit system, leaving thousands without a way to work', *Associated Press*, 9 August 2007.

37 *'Buried beneath our feet'* – NRDC, op. cit., p.v.

– *'The depressing attitude'* – House of Commons Environmental Audit Committee, 'Corporate Environmental Crime, Second Report of Session 2004–5', London: The Stationery Office Ltd., 2005, p.3.

– *A £2 billion interceptor stormwater tunnel* – 'New tunnel to give London a 21st century River Thames', Department for Environment, Food and Rural Affairs, news release, 22 March 2007.

38 *'Frighten a lady into asterisks'* – Mayhew, 1851, op. cit., p.431.

41 *A Haitian immigrant* – Marie Brenner, 'Incident in the 70th Precinct,' *Vanity Fair*, December 1997.

– *The contribution of sewer-workers* – Recently, scientists have been using 'sewage epidemiology' to estimate levels of drug use. At Italy's Istituto di Ricerche Farmacologiche Mario Negri, researchers tested sewage samples from Milan, London and Lugano, Switzerland, for cocaine, opiates, cannabis and amphetamines. Findings included the fact that 40,000 doses of cocaine are taken daily in Milan, far above the previous estimate of 15,000. Though sewage epidemiology can be tricky – it requires assuming the size of an average dose to how many people pee to how far the sewage has travelled – the technique is thought to be more reliable than surveys, the usual tools for calculating drug abuse. People lie; their sewage doesn't. Ettore Zuccato et al, 'Estimating Community Drug Abuse by Wastewater Analysis', *Environmental Health Perspectives EHP-in-press*, May 1, 2008, posted online at http://www.ehponline.org/docs/2008/11022/abstract.html

– *$5.3 million in damages* – The full settlement of $8.7 million was reduced to $5.3 million after legal fees: 'Victim, Civil Rights leader look back on infamous NYPD torture case', *International Herald Tribune*, 9 August 2007.

42 *Natural percolation of water* – Two years of monitoring Seattle's SEA streets concluded that the volume of storm-water leaving the street had been reduced by ninety-nine per cent. More information from http://www.seattle.gov. See also Naomi Lubick, 'Using nature's design to stem urban storm-water problems', *Environmental Science and Technology*, vol. 40, no. 19, 1 October 2006, pp.5832–3.

– *Twenty-two Hyde Parks* – Gravel, porous bricks or paving are exempt from the new requirement. Rebecca Smithers, 'New rules for front gardens to fight floods', *Guardian*, 8 February 2008.

– *Roosevelt said* – Not only did Roosevelt dare to bring up the topic of sewage and water conservation before the businessmen of Buffalo, some of whom had interests in Niagara water power, but he did it at a 6 a.m. breakfast appointment. 'Roosevelt Stirs Middle West Again; Crowds from Buffalo to Chicago Cheer Attacks on Corporate and Private Dishonesty', *New York Times*, 26 August 1910.

– *The Art of Drainage* – Joseph Bazalgette may be relatively obscure, but at least he has a Wikipedia entry; George. E. Waring, an equally innovative, daring and determined engineer, does not. George E. Waring, *Draining for Profit, and Draining for Health*, originally published New York: E. Judd & Co., 1867; Ebook 19465, Project Gutenberg, 4 October 2006, p.230.

Chapter Two: The Robo-Toilet Revolution

45 *The best lavatories* – Adam Hart-Davis, *Thunder, Flush and Thomas Crapper*, London: Michael O'Mara Books, 1997, p.26.

46 *Two dozen or so a year* – Ibid., p.48.

– *Fire and the wheel* – The top five inventions of all time were, in order, the toilet, the computer, the printing press, fire and the wheel. *Focus*, May 1997, pp.68–76.

49 *A presence in 16 countries* – All TOTO figures taken from TOTO annual report 2006, available from http://www.toto.co.jp/en/

51 *A tiny proportion of the country was sewered* – Even in 1961, only six per cent of the population was sewered. *Making Great Breakthroughs: All about the Sewage Works in Japan*, Tokyo: Japan Sewage Works Association, 2002, p.7.

– *More Japanese were sitting than squatting* – Takushi Ohno, 'Bowled over by toilet's evolution into high-tech gizmo', *Asahi Shimbun*, 8 December 2005.

– *Toilet Ho!* – Heisei Kawaya Kenkyukai and Minoru Sato, *Toilet de HO*, Tokyo: TOTO Publishing, 1995.

52 *'A vicious libel'* – Alexander Kira, *The Bathroom*, New York: The Viking Press, 1966, 1976, p.25.

– *She had 'never seen the Prophet'* – Abdul Fattah Al-Husseini Al-Sheikh, 'Water and Sanitation in Islam', Cairo: WHO Regional Office for the Eastern Mediterranean, 1996.

53 *'Their faecally stained pants'* – Dr Cameron surveyed 940 men in Oxfordshire, of whom 43.8 per cent showed faecal contamination of their underpants. 8.6 per cent were not accustomed to wearing underpants at all. Dr Cameron was diligent in reporting his findings to the participants, saying, 'I decided there was no point in this Survey unless I mentioned to those involved that they showed a lack of hygiene in this particular regard. It was naturally a delicate matter to handle, and the answers varied between a truculent "mind your own business" attitude and a host of excuses such as sweating, long journeys, bicycle riding, "nervous diarrhoea" and haemorrhoids.' Under 'Possible Remedial Measures', Dr Cameron suggested that laundries adopt a less tolerant attitude; hygiene education; a preparedness to tackle an unsavoury matter with the offenders whenever this was brought to light; and more bidets. J.A. Cameron, 'A Particular Problem Regarding Personal Cleanliness', *Public Health*, vol. 76, 1972, pp.173–7.

– *'Grazing the wings of butterflies'* – Jun'ichiro Tanizaki, 'One thing and another on the privy', *Postcolonial Studies*, vol.5, no.2 (2002), pp.147–51.

– *Kin no Unko (Golden Poo)* – Alice Gordenker, 'So What the Heck is that?' *Japan Times*, 20 March 2007.

59 *twenty-three per cent of Japanese houses* – Kazuo Mikami, 'Promising an "Evolution" in Toilet Culture', *Japan Times*, 14 May 2007.

61 *'Complex and ridiculous thrones'* – Alan Watts, *In My Own Way: An autobiography, 1915–1965*, London, Jonathan Cape, 1973, pp.33–5.

– *'The American toilet'* – U.S. EPA, 'How to conserve water and use it effectively,' from http://www.epa.gov/nps/chap3.html

62 *One cross-border black marketeer* – CNN transcript 00052100V55, 21 May 2000.

63 *'Get the federal government out of the bathroom'* – Joe Knollenberg, 'Federal Government has no business in your bathroom', *Environment & Climate News*, 1 December 1999.

– *'Plumber's smudge'* – Wallace Reyburn, *Flushed with Pride: The Story of Thomas Crapper*, London: Macdonald and Company, 1969, p.17.

64 *Ranked first, second and third* – NAHB Research Center, Water Closet Performance Testing, September 2002, www.cuwcc.org/toilet_fixtures/NAHB_ToiletReport_Sept-2002.pdf

65 *An average bowel movement* – J.B. Wyman, K.W. Heaton, A.P. Manning and A.C. Wicks, 'Variability of colonic function in healthy subjects', *Gut*, 1978; 19, pp.146–50.

66 *Veritec's MaP* – 'Maximum Performance (MaP) Testing of Popular Toilet Models' (MaP) reports available from http://www.veritec.ca

67 *'Clean is Happy' campaign* – http://www.cleanishappy.com

– *'The most significant innovation for personal hygiene'* – American Bidet Company, via http://www.bidet.com

68 *Now it's ten per cent* – Adam Sage, 'French home keeps children away', *The Times*, 16 March 2005.

– *'The little bathroom for the feet'* – Harvey A. Levenstein, *We'll Always Have Paris: American Tourists in France since 1930*, Chicago: University of Chicago Press, 2004, p.31 quoting Jan Gelb's diary, 13 July 1929.

– *'Highly charged emotionally'* – Kira, op. cit., p.23.

– *French, and therefore louche* – Personal communication with Harvey Molotch, August 2007. See also Harvey Molotch, *Where Stuff Comes From: How Toasters, Toilets, Cars and Computers Came to Be As They Are*, New York: Routledge, 2005.

– *Its bidet model Carmen* – Kira, op. cit., p.23.

– *'A freakish bathroom custom'* – 'On the Down-low,' Harper's Magazine, August 2005.

69 *'Close your eyes and seek God'* – Dareh Gregorian and Kathianne Boniello, 'The Bums' Rush', *New York Post*, 31 July 2007.

– *'Ask for Hakle'* – *Verlangen Sie eine Rolle Hakle®, dann brauchen Sie nicht Toilettenpapier zu sagen!* http://www.hakle.de/about_history.asp

– *Toilet paper industry is worth $15–$20 billion* – Conal Walsh, 'Put through the mill by US toilet tissue titans', *Observer*, 20 July 2003.

– *Fifty-seven sheets a day* – Personal communication with Celeste Kuta, Charmin, February 2008.

– *'Love your bum'* – D&AD, Velvet case study, 2005, p.1, available from www.dandad.org/inspiration/creativityworks/5/pdf/Velvet.pdf
– *Second most complained-about ad* – Advertising Standards Authority annual report 2003, p.5.
70 *Sales of Rollwipes were dismal* – Irina Barbalova, 'Wet toilet tissue yet to win over consumers', *Euromonitor*, 29 May 2003.
71 *American toilet market leader Kohler* – 'Complete your bathroom with a high-tech Kohler toilet seat', Kohler, news release, 20 June 2007.
– *'A field of daisies'* – Barry Sonnenfeld, 'Tucks can change your life', *Esquire*, 1 June 2004.

Chapter Three: 2.6 Billion

74 *The rather bigger group of the same name* – 'In praise of public loos', *Guardian*, 31 July 2007.
76 *One in three of the world's people* – Personal communication with Pete Kolsky, November 2007.
– *'A deficit of action'* – UNDP, op. cit., p.70.
– *Twenty-three UN agencies* – Ibid., p.8.
– *No act of terrorism* – Ibid., p.3.
77 *Winners' revenue allegedly doubles* – Jehangir S. Pocha, 'Just Flush with Pride', *Boston Globe*, 26 November 2004.
78 *A staggering eighty-seven per cent* – Nellie Bristol, 'Mechai Viravaidya: Thailand's "Condom King"', *Lancet*, vol. 371, 2008, no. 9607, p.109. A 2006 report estimated that 19,500 new infections will occur in 2004, compared with 143,000 new infections in 1990. Without the prevention campaigns, 7.7 million more people would have been infected, or fourteen times more than is now the case. Ana Revenga *et al.*, 'The Economics of Effective AIDS Treatment: Evaluating Policy Options for Thailand', Washington, DC: The World Bank, 2006, pp.xxii, xxiii.
– *Eight World Toilet events* – Personal communication with Jack Sim, October 2007.
79 *Stabbed by a monk* – Monestier, op. cit., p.44.
– *Hiding in his privy* – The king had apparently blocked up his only chance of escape – the hole through which the privy contents were supposed to fall to the ground below – after he had been playing a ball-game and lost too many

balls down the hole. He was finally stabbed sixteen times after 'that odious and false traitour, sir Robert Grame [descended] down also into the privy to the king, with an horribill and mortall weapon in his hand'. John Shirley, *Life and Death of King James I of Scotland* (ed. Joseph Stevenson), Edinburgh: Maitland Club, 1837, accessed from the National Libraries of Scotland at http://www.nls.uk/scotlandspages/timeline/1437.html

81 *Celebrities who do charity work for water* – 'Water wins the celebrity Oscar – sanitation stuck on the B-list', International Water and Sanitation Council/IRC, news release, 23 March 2007.

– *'A smile on my face'* – WSSCC, *Listening*, p.23.

82 *An average $7 return* – Guy Hutton, Laurence Haller and Jamie Bartram, 'Global cost-benefit analysis of water supply and sanitation interventions', *Journal of Water and Health*, May 2007, pp.481–502.

– *£54,000 a year* – David Redhouse, Paul Roberts and Rehema Tukai, 'Every one's a winner? Economic valuation of water projects', Water Aid Discussion Paper, 2004, p.3, quoting Tristram Hunt, *Building Jerusalem: The Rise and Fall of the Victorian City*, London: Weidenfeld & Nicholson, 2004, p.259.

– *It would save $660 billion* – Hutton, Haller and Bartram, op. cit., and personal communication with Guy Hutton, March 2008.

– *Losses from agricultural revenue* – John Oldfield, 'Community-based approaches to Water and Sanitation: A Survey of Best, Worst and Emerging Practices', draft paper for the Navigating Peace Initiative of the Woodrow Wilson International Center for Scholars' Environmental Change and Security Program, p.13.

82 *Averting a child death* – Sandy Cairncross and Vivian Valdmanis, 'Water Supply and Sanitation', in *Disease Control Priorities in Developing Countries*, 2nd Edition, eds. D.T. Jamison, G. Alleyne, J.G. Breman, M. Claeson, D.B. Evans, Prabhat Jha et al, Washington, D.C.: The World Bank, 2006, p.776–78.

– *47 times more on its military budget* – United Nations Development Programme, *op. cit.*, p.62.

84 *Blowing up some electricity pylons* – Electricity pylons from Kasrils' application for amnesty from South Africa's Truth and Reconciliation Commission, available at http://www.doj.gov.za/trc/decisions/2001/ac21168.htm; other biographical information from the African National Congress website at http://www.anc.org.za/people/kasrils.html; and from personal communication with Lorna Michaels at South Africa's Ministry of Intelligence. For more on Kasrils, including his career as the poet ANC Khumalo, see his

autobiography *Armed and Dangerous: My Undercover Struggle under Apartheid*, Johannesburg: Jonathan Ball Publishers, 1993. Kasrils' senior sanitation adviser during his time as Minister for Toilets was Denis Goldberg, who was convicted along with Nelson Mandela at the Rivonia Trial in 1964, and spent twenty-two years in prison. He's now particularly interested in latrine emptying by enzymes.

85 *'Victory with its pants down'* – 'The great anatomist Sidrac', in conversation with doctors Goudman and Grou, as told by Voltaire, believed strongly in the power of constipation, telling his listeners that Oliver Cromwell had not visited the necessary house for eight days before he cut off King Charles's head. Equally, Sidrac stated, the Duke of Guise le Balafré was often advised to avoid King Henri III on a winter's day when the north-east wind was blowing, because the king would be attempting to alleviate his constipation on his close-stool. François Voltaire, *Candide, Zadig and Selected Stories* (translated by Donald Murdoch Frame), New York: Signet Classic, 2001, p.344.

– *The colour that flies liked least* – Monestier, op. cit., p.179.

– *The excreta is deadly* – According to Charles Heyman of *Jane's Defence Weekly*, *pungis* are still used to guard opium poppy fields in south-east Asia. In Colombia, meanwhile, the revolutionary group ELN has used bombs containing clay mixed with human faeces, to increase the risk of infecting wounds. Personal communication with Charles Heyman. Mariano C. Bartolome and Maria Jose Espona, 'Chemical and Biological Terrorism in Latin America: The Revolutionary Armed Forces of Colombia', *ASA Newsletter*, Applied Science and Analysis, Inc., 2003, vol. 5, no. 1998. Also: E. Lesho, D. Dorsey and D. Bunner, 'Feces, dead horses, and fleas – Evolution of the hostile use of biological agents', *Western Journal of Medicine*, 1998, 168, p.512.

87 *'Wow. Bang'* – Personal communication with Stephen Turner, WaterAid, March 2007.

– *HIV, safe water, malaria or nutrition* – Arno Rosemarin, 'EcoSanRes Programme – Phase Two 2006–2010', Eschborn: DWA, Hennef and GTZ, 2006, p.9.

88 *Ninety-two have never counted them* – 'The eight commandments', *Economist*, 5 July 2007.

– *Jack and Jill* – WSSCC has recently updated its campaign. New images include a blindfolded man standing in front of a firing squad wall, but his executioners are holding glasses of water, not guns. The slogan reads 'Dirty Water Kills'. Another shows sperm-shaped water droplets and the slogan 'In some countries women risk rape by collecting water'. All campaign images are available at http://www.wsscc.org

89 *The Blair latrine* – Dr Peter Morgan, 'An Ecological Approach to Sanitation in Africa: A Compilation of Experiences', available from http://www.ecosanres.org/PM_Report.htm

– *179 flies a day* – Andy Robinson, 'VIP Latrines in Zimbabwe: From Local Innovation to Global Solution', Nairobi, Kenya: Water and Sanitation Program – Africa Region, 2002, p.2.

96 *School enrolment increased* – UNICEF, 'Children and Water, Sanitation and Hygiene: The Evidence', Human Development Report Office Occasional Paper No. 50, New York: UNDP, 2006, p.3.

Chapter Four: Going to the Sulabh

102 *'Longest surviving social hierarchy'* – Smita Narula, *Broken People: Caste Violence against India's 'Untouchables'*, New York: Human Rights Watch, 1999, p.24.

– *India's British rulers* – The British created official posts of manual scavengers and used scavengers to clean army cantonments and municipalities. Dry toilets abounded in all British-run environments. Gita Ramaswamy writes, 'This is not to say the British invented caste or manual scavengers; rather they intervened specifically to institutionalise it.' Gita Ramaswamy, *India Stinking: Manual Scavengers in Andhra Pradesh and Their Work*, Pondicherry: Navayana, 2005, p.6.

– *Between 400,000 and 1.2 million manual scavengers* – Ramaswamy quotes figures from the Ministry for Social Justice and Empowerment, which calculated in 2002–3 that there were 676,000 manual scavengers in India. In his foreword to *India Stinking*, Safai Karamchari Andolan leader Bezwada Wilson puts the total at 1.3 million. Ibid., p.vi, ix.

103 *Ten million dry latrines* – WSSCC, *Listening*, p.36.

– *Manual scavenging illegal* – Narula, op. cit., p.149.

– *The Constitution of India* – Article 17 reads in full – '"Untouchability" is abolished and its practice in any form is forbidden. The enforcement of any disability arising out of "Untouchability" shall be an offence punishable in accordance with law.' Constitution of India, 1949, available from the Government of India's Ministry for Law and Justice at http://lawmin.nic.in/coi.htm

– *Three men died of asphyxiation* – 22,237 Dalits die doing sanitation work every year. S. Anand, 'Life Inside a Black Hole', *Tehelka*, 8 December 2007.

105 *'Loses arm as a result'* – Stephanie Barbour, Tiasha Palikovic, Jeena Shah, and Smita Narula, 'Caste Discrimination against Dalits or So-called Untouchables in India, Submission to the Committee on the Elimination of Racial Discrimination', New York: Center for Human Rights and Global Justice, New York University/Human Rights Watch, 2007, p.40.

– *Three Dalit women were raped* – National Campaign on Dalit Human Rights, 'Alternate Report to the Joint 15th to 19th Periodic Report to the State Party (Republic of India) to the Committee for the Elimination of Racial Discrimination', New Delhi: National Campaign on Dalit Human Rights, 2006, p.2.

– *Sit apart from other children* – Ibid., p.10.

106 *Spade, Black, Dung, Horse* – Ramaswamy, op. cit., p.17.

– *Fifteen duties for slaves* – Ibid., p.5.

– *Baiting of cesspit-emptiers* – The French royal ordinance of 1350 authorized labourers from all trades and guilds to become cesspit emptiers if they chose, and stipulated that 'whosoever insults them or does them violence will be fined'. Monestier, op. cit., p.100.

107 *In a northwesterly direction* – The Essenes, widely believed to have written the Dead Sea Scrolls, suffered for their good hygiene, according to recent research in the Scroll site of Qumran in the West Bank. Because Qumran's inhabitants designated a specified place as a latrine and buried their waste, rather than following the Bedouin custom of open defecation (which meant faeces dried out in the sun), they tramped faecal pathogens back into their camp. This could explain why only six per cent of male corpses found in Qumran had survived beyond the age of forty, while in first-century Jericho, nine miles away, the rate was forty-nine per cent. Alan Boyle, 'Toilet tied to tale of Dead Sea Scrolls', MSNBC, 15 November 2006.

– *'Smell thine own dung'* – J.G. Bourke, *Scatologic Rites of All Nations: A Dissertation upon the Employment of Excrementitious Remedial Agents in Religion, Therapeutics, Divination, Witchcraft, Love-Philters, etc., in all Parts of the Globe*, Washington, D.C.: W.H. Lowdermilk & Co., 1891, p.143.

– *Fire an arrow* – John Stackhouse, 'Clean Revolution begins with Outhouse', *Toronto Globe & Mail*, 19 December 1994.

– *15 feet away from habitation* – Sulabh International, 'Sulabh International Museum of Toilets', undated, p.13.

– *A rough stick* – Vitta Khandaka, *Collection of Duties*, translated by Thanissaro

Bhikkhu, from http://www.accesstoinsight.org/tipitaka/vin/cv/cv.08x.than.html

108 *The current Chief Justice* – Chief Justice K.G. Balakrishnan, known to his friends as Bala, is the first Dalit to hold the post. Narenda Jadhav, chief economist of India's central bank, is another Dalit success story. They are the exceptions. Amelia Gentleman, 'India's untouchable millionaire', *Observer*, 6 May 2007; Balakrishnan takes over as new C.J.I., *The Hindu*, 15 January 2007.

– *Won't marry outside their own caste* – Sagarika Ghose, 'The cult of the sex goddess', *Guardian*, 14 August 2007.

– *When I first wrote about manual scavengers* – Rose George, 'These Indian women are forced to get way too intimate with their bosses. And not in a good way', *Jane*, August 2003.

– *The cheapest cleaning option* – In the latest railway budget, Indian Railways did however allocate £512 million to install three types of toilet – controlled discharge (to be released only when the train is going more than eighteen miles an hour), biodegradable and vacuum-retention – in 36,000 railway coaches. The reason for this largesse was that the 300,000 litres of human excreta gushing from India's trains are rapidly corroding the rails. No mention was made of scavengers. Rahul Bedi, 'India to invest millions in "green" train toilets', *Daily Telegraph*, 4 March 2008.

– *A high court in Nizamabad* – Bezwada Wilson, 'He cleaned up the system', *Tehelka*, 7 October 2006.

109 *The total eradication of manual scavenging* – 'India plans new system to rehabilitate scavengers', *Deutsche Presse-Agentur*, 30 January 2007.

110 *Their substance was subject enough* – Adrian Searle, 'Absolute Excrement', *Guardian*, 4 December 2007.

111 *He publicly cleaned his own* – 'There was no limit to insanitation. [...] There were only a few latrines, and the recollection of their stink still oppresses me. I pointed it out to the volunteers. They said point blank, "That is not our work, it is the scavenger's work." I asked for a broom. The man stared at me in wonder. I procured one and cleaned the latrine. But that was for myself. The rush was so great, and the latrines so few, that they needed frequent cleaning, but that was more than I could do. [...] And the others did not seem to mind the stench and the dirt.' M.K. Gandhi, *An Autobiography or The Story of My Experiments with Truth*, Ahmedabad: Navajivan Publishing House, 1927, reprinted 2006, pp.206–7.

– *'Evacuation is as necessary as eating'* – 'Everyone must be his own scavenger. [...]

I have felt for years that there must be something radically wrong, where scavenging has been made the concern of separate class in society.' M.K. Gandhi, *Harijan*, Vol. XVIII, 30 October 1954.

– *Inequality was the soul of Hinduism*: B.R. Ambedkar, *Writings and Speeches*, Bombay: Government of Maharashtra, 1987, vol. 3, p.66.

112 *'Quite an achievement of the imagination'* – Smith, op. cit., p.29.

– *'The utmost violation of human rights'* – Bindeshwar Pathak, 'Operation, Impact and Financing of Sulabh', Occasional Paper for the Human Development Report 2006, p.4.

113 *232 of India's 5233 towns* – Ibid.

– *The kind you get from chronic diarrhoea* – For more on Delhi's sewage and the dying Yamuna, see *Faecal Attraction: The Political Economy of Defecation*, a film by Pradip Saha, available from http://csestore.cse.org.in

– *A pioneering composter* – Gandhi gets most of the credit, but India has a proud history of pioneering sanitation activists and composters, including Appa Saheb Patwardhan, whose name lives on in a foundation set up by another pioneering sanitation activist, Dr S.V. Mapuskar, who has installed seventy-five human waste biogas digesters in his home town of Dehu, near Pune, making it unique. S.P. Singh, *Sulabh Sanitation Movement: Vision-2000 plus*, New Delhi: Sulabh International, 2005, p.306.

114 *Poor people in Uganda* – United Nations Development Programme, op. cit., p.51.

– *The eminence of Faraday* – Singh, op. cit., p.iii.

119 *Well-known businessmen readily agree* – Pathak, 2006, op. cit., p.11.

121 *Half a million Indians* – WSSCC, *Listening*, 2004, p.37.

– *A candle in the dark* – Swiss Agency for Development and Co-operation, 'Sanitation is a business', Bern: SDC, 2004, p.18.

Chapter Five: China's Biogas Boom

124 *Their soils turned to dust* – David Montgomery, *Dirt: the erosion of civilizations*, Berkeley and Los Angeles: University of California Press, 2007, p.181. The agricultural scientist Franklin Hiram King, who spent nine months travelling through China and Japan in 1909, was as impressed by Asia's sensible use of all waste as he was scornful of the West's waste of it. He concluded that 'the people of the United States and of Europe are pouring into the sea, lakes or

rivers and into the underground waters from 5,794,300 to 12,000,000 pounds of nitrogen; 1,881,900 to 4,151,000 pounds of potassium, and 777,200 to 3,057,600 pounds of phosphorus per million of adult population annually, and this waste we esteem one of the great achievements of our civilisation'. F.H. King, *Farmers of Forty Centuries*, originally published 1911, Whitefish, Montana: Kessinger Publishing, 2004, p.63.

– *'Threw her into the toilet to die'* – *Cesuo wenhua manlun* (General essays on toilet culture), eds. Feng Shufei, Zhang Yiguo and Zhang Dongsu, Shanghai: Tongji University Press, 2005, pp.49–50.

125 *'Stinky shit egg'* – Shanghai nightsoil-collector-turned-competitive-bicyclist Chen Qiaozhu told of her experiences 'before she came out of the toilet to claim victory. Passers-by would see her and "pooh-pooh her, plug their noses, and spit," to which Chen would fiercely reply, "look! If we stopped doing this work for just three days, Shanghai would turn into Stinkytown!" Nightsoil collector heroes were politically useful because "only in a truly liberated society could a collector and carrier of human waste be given the honor of making an address in the hallowed Great Hall of the People."' Andrew Morris, '"Fight for Fertiliser!" Excrement, Public Health and Mobilization in New China', *Journal of Unconventional History*, vol. 6, no. 3, Spring 1995, pp.52, 64, 69.

– *Seas of Shit, Mountains of Fertiliser* – Ibid., pp.61–2.

126 *15.4 million rural households* – Personal communication with Institute of Biogas (BIOMA), Chengdu, China, August 2006.

127 *The excreta of genocidal murderers* – Biogas digesters are now installed in six Rwandan prisons. Using biogas instead of wood for energy has cut wood use by sixty per cent and saved £1 million in fuel costs. The odour-free compost that results from biogas has also done wonders for the prison gardens. 'Rwanda award for "sewage" cooking', BBC News, 30 June 2005.

130 *A sizeable international prize* – Ashden Awards for Sustainable Energy, 'Ashden Awards Case Study, Shaanxi Mothers', 2006, http://www.ashdenawards.org/winners/shaanxi

– *It increased vegetable yields* – J. Paul Henderson, 'Anaerobic Digestion in Rural China', City Farmer, Canada's Office of Urban Agriculture, available from http://www.cityfarmer.org/biogasPaul.html

– *270 square metres of land* – Heinz-Peter Mang, presentation at the World Toilet Summit, Moscow, September 2006.

– *Six hours' worth of a 60–100 watt bulb* – L.A. Kristoferson and V. Bokalders,

Renewable Energy Technologies: their application in developing countries, London: ITDG Publishing, 1991, p. 108.

– *2100 kilograms of wood* – Henderson, op. cit.

– *Count Alessandro Volta* – 'The appearance of flickering lights emerging from below the surface of swamps was noted by Pliny and Van Helmont recorded the emanation of an inflammable gas from decaying organic matter in the seventeenth century. Volta is generally recognised as putting methane digestion on a scientific footing.' Uri Marchaim, 'Biogas Processes for Sustainable Development', Rome: Food and Agriculture Organization, 1992, Chapter 4.

131 *A leper colony in Bombay* – Opinions differ as to whether the Ackworth Leper Colony in Matunga, Bombay installed its pioneering biogas digester in 1859 or 1896. For 1859, see Paul Harris' informative biogas pages at http://www.adelaide.edu.au/biogas. For 1896, see Marchaim, op. cit.

132 *30 kilograms of faeces daily* – Henderson, op. cit.

– *Strike for three days* – David G. Strand, 'Feuds, Fights and Factions: Group Politics in 1920s Beijing', *Modern China*, vol. 11, no. 4, October 1985, pp.425–7.

133 *Sister Ah Gui . . . Shit Queen* – Hanchao Lu, *Beyond the Neon Lights: Everyday Shanghai in the Early Twentieth Century*, Berkeley: University of California Press, 1999, p.194. For a fuller account of China's *fen* business, see F. H. King, op. cit., p.63.

– *A million . . . 'honey bucket[s]'* – Lu, op. cit., pp.191–3.

– *The nightsoil man's wooden cart* – Emily Prager, 'Settling down in a city in motion', *New York Times*, 19 July 2007.

– *If current rates of exploitation continue* – Ingrid Steen, 'Phosphorus availability in the 21st century: Management of a non-renewable resource', *Phosphorus & Potassium*, 217, September–October, 1998, p.12.

135 *The problem of sanitation was solved* – 'Health and Sanitation in Communist China', U.S. Joint Publications Research Service, Report 96, 17 December 1957, p.5.

139 *Cost is the biggest obstacle* – Marion W. Jenkins, 'Who Buys Latrines, Where and Why?' (Field note/WSP), Nairobi, Kenya: Water and Sanitation Program – African Region, 2004, p.9.

140 *Fossilized Peruvian dung* – Raúl Patrucco, Raúl Tello and Duccio Bonavia, 'Parasitological Studies of Coprolites of Pre-Hispanic Peruvian Populations', *Current Anthropology*, vol. 24, no. 3, June 1983, pp.393–4.

– *'No such thing as a total removal'* – Matthias Gustavsson, 'Biogas – Solution in search of its problem', Göteborg University Human Ecology Reports Series 1, March 2000, p.16.

141 *Over two-thirds of Chinese* – By 2030, a billion Chinese will live in cities. Jonathan Woetzel, Janamitra Devan, Luke Jordan, Stefano Negri and Diana Farrell, 'Preparing for China's Urban Billion', Shanghai: McKinsey Global Institute, 2008, p.14.

– *Sufficient dung* – Jungfen (Jim) Zhang and Kirk R. Smith, 'Household Air Pollution from Coal and Biomass Fuels in China: Measurements, Impacts and Interventions', *Environmental Health Perspectives*, 115 (2007), p.854.

Chapter Six: A Public Necessity

146 *'Aural privacy'* – Government standards recommend that background noise be increased to a level of 55 db (A) – a decibel level at which loud speech can be understood – to aid embarrassed students. The same report also suggests that urinals be avoided, because their use by boys at puberty is 'problematic'. Trough-type urinals can contribute to shy bladder syndrome. Department for Education and Skills, 'Standard Specifications, Layouts and Dimensions 3: Toilets in Schools', London: Department for Education and Skills, 2007, p.9.

147 *'Civil inattention'* – Kira, op. cit., p.204.

– *'Pearls and rubies'* – Norbert Elias, *The Civilizing Process*, Oxford: Blackwell Publishing, 2000, p.123.

148 *Leaping stags* – Dr Paola Chini, '*Latrina Romana in Via Garibaldi*', Comune di Roma Assesorato alle Politiche Culturali Sovraintendenza ai Beni Culturali, available from http://www2.comune.roma.it/monumentiantichi

– *'It costs two sous to do it'* – *'Chacun sait ce qu'il a à faire et il faut payer deux sous.'* J.F.C. Blanvillain, *Le Pariseum*, 1807, quoted in Monestier, op. cit., p.147.

– *Mr Tinkler* – Tinkler won his case, as the judges decided that Wandsworth Board of Works had exceeded its statutory requirement by forcibly installing water closets. John Walter de Longueville Gifford, *Reports of cases adjudged in the High Court of Chancery by the Vice-Chancellor Sir John Stuart*, London: Wildy & Sons, 1860, pp.412–420.

149 *Private-public facilities installed by George Jennings* – Lawrence Wright, *Clean and Decent: the fascinating history of the bathroom and the water closet, and of sundry habits, fashions and accessories of the toilet, principally in Great Britain,*

France and America, London: Routledge & Kegal Paul, 1960, p.200.

– *Paris Métro's new Line 1* – These magnificent art-deco facilities also had a steel staircase decorated with mosaics, mahogany doors and fittings and copper pipes. Monestier, op. cit., p.154.

– *In London forty-seven per cent of public toilets* – 'Sassy Britannia will spend a giant penny to honour pioneering public toilet', Royal Society of Chemistry, news release, 13 August 2002.

– *Decreased in number by forty per cent* – Clara Greed, 'The role of the public toilet: Pathogen transmitter or health facilitator?' *Building Service Engineering Research and Technology*, May 2006, vol. 27, pp.127–39.

– *'Run the gamut'* – The Council of the City of New York, 'Toilet Trauma: A Survey of Public Restrooms in New York City', August 2001, p.4.

150 *'Public toilets! Has it come to this?'* – Roger Kimball, Armavirumque (*New Criterion* weblog), 30 May 2005, http://www.newcriterion.com/weblog/2005/05/where-is-hercules-when-you-need-him.html

151 *'The acceptability of eliminating these substances'* – Pinker, op. cit., p.344.

– *'Simon Norris, The Disabled Toilet'* – Maev Kennedy, 'Seaside squat is a room with a loo', *Guardian*, 9 April 2007.

– *Overnight accommodation by Polish migrant workers* – The desirable residential toilets in Stamford Hill were later fitted with new locks to deter squatters. Matthew Hickley and Laura Roberts, 'The superloo where Polish migrants are fighting to spend the night for 20p', *Daily Mail*, 3 May 2007.

152 *Ninety-five per cent of Britons* – 'Should we splash out to spend a penny?' ENCAMS, news release, 19 July 2006.

153 *Forced to go on strike* – Monestier, op. cit., p.26.

154 *'The writing on the shithouse walls'* – William Rivers Pitt, 'The Writing on the Latrine Walls', Truthout.org, 9 August 2004, accessed from http://www.truthout.org/cgi-bin/artman/exec/view.cgi/50/5656

155 *That might 'frighten the horses'* – 'I am a liberal and I believe in freedom and responsibility. In a free country, consenting adults should be able to do as they please in private, so long as they do not frighten horses. However, I should put sex in public toilets into the category of frightening the horses – or at any rate, the children.' Hansard HL (House of Lords), Vol. 648, col. 584, 19 May 2003.

– *'Anti-pipi' wall* – The wall is gently angled so that the jet of urine sprays back onto the urinator. 'It is a case of the hoser being hosed,' according to Emile

Vanderpooten, the municipal architect who devised it – Henry Samuel, 'Paris mayor moves to stop public urinating', *Daily Telegraph*, 29 October 2007.

– *The volumes of animal excreta* – The Mairie de Paris estimates that the city's 20,000 dogs produce 16 tonnes of 'canine pollution' daily, which cost 11 million euros a year to clean up, and which cause 650 accidents a year. Mairie de Paris, '*Bien vivre avec les animaux à Paris*', via www.paris.fr

156 *Three million visitors* – 'Beijing airport to launch emergency alert mechanism for Olympics', *Xinhua*, 25 June 2005.

– *Five minutes' walking distance* – Ibid.

157 *Ninety-six specific rules* – Morris, op. cit., p.53.

– *'Show mercy to the slender grass'* – 'Beijing getting rid of badly translate [sic] signs', *China Daily*, 27 February 2007.

– *Guo Zhangqi [the fly-catcher]* – 'Two farmers' desire to wipe out flies in Beijing', *Xinhua*, 6 June 2003.

159 *'The bladder leash'* – Help the Aged, 'Nowhere to Go: Public Toilet Provision in the UK', 2007, p.5.

– *A national toilet map* – The website of Australia's National Public Toilet map allows visitors to check location and opening times of toilets. New South Wales has the most toilets with 4446; Northern Territory has only 198. http://www.toiletmap.gov.au

– *Satlav* – Westminster Council's Satlav service covers 8.5 square miles of central London. Users send an SMS to 80097 – at 25p per request – and receive in return a list of the nearest facilities and opening hours. James Meikle, 'Satlav to end problem of users being caught short', *Guardian*, 30 November 2007.

160 *'Potty parity'* – During a television interview on CNN's Crosstalk in 2002, John Banzhaf mentioned the use of funnels for women to urinate through (which can be easily bought these days in Swedish pharmacies). Co-anchor Robert Novak responded to this information by saying, 'This show is becoming X-rated.' CNN Crossfire, 9 August 2002, http://transcripts.cnn.com/TRANSCRIPTS/0208/09/cf.00.html

– *Ninety seconds to urinate* – Alexander Kira calculates the mean occupancy of men at a urinal to be forty-five seconds. The Malaysian Standard prefers thirty-five seconds. Malaysian Standard Public Toilets MS 2015: Part 1: 2006, p.31. See also Clara Greed, *Inclusive Urban Design: Public Toilets*, Oxford: Architectural Press, 2003, p.8.

– *'Four urinals, gilt mirrors'* – Sarah A. Moore, 'Facility Hostility? Sex Discrimination and Women's Restrooms in the Workplace', 36 *Georgia Law Review*, 599, 2002, p.599.

161 *'A keypad code'* – Margaret Talev, '"Potty Parity"' surfaces in House of Representatives', *Columbian*, December 24, 2006.

– *More sharing than many people feel comfortable with* – Kira, op. cit., pp.201–2.

162 *Stickers in the bowl* – http://blog.wired.com/gadgets/2006/09/bowl_light_lets.html

– *'Stand-peeing'* – Kate Connolly, 'German men told they can no longer stand and deliver', *Daily Telegraph*, 17 August 2004.

– *Klaus Schwerma* – Klaus Schwerma, *Stehnpinkeln: Die letze Bastion der Männlichkeit?* Bielefeld: Kleine Verlag, 2000.

– *Herodotus reported* – 'Women urinate standing up, while men do so squatting. They relieve themselves indoors, while they eat outside on the streets. The reason for this, they say, is that things that are embarrassing but unavoidable should be done in private, while things which are not embarrassing should be done in the open.' Herodotus, *The Histories* (translated by Robin Waterfield), Oxford: Oxford University Press, 1998, p.109. (In my 1996 Wordsworth Classics edition, though supposedly unabridged, all references to urination in this section had been removed.)

163 *She-Pee female urinals* – Glastonbury Bog Blog, WaterAid, 27–9 June 2007, http://www.wateraid.org/uk/get_involved/events/festival_events/glastonbury/5591.asp

– *Twenty carloads of people* – Patrick Barkham, 'Bog Standards', *Guardian*, 24 September 2001.

164 *Foot-tapping* – standing in plastic bags: 'Craig resigns over airport sex sting', *Associated Press*, 2 September 2007; and Christopher Hitchens, 'So Many Men's Rooms, So Little Time', *Slate*, 3 September 2007.

165 *'It would be a good idea'* – Mohandas K. Gandhi (attributed), *Chambers Dictionary of Quotations*, Edinburgh: Chambers Harrap Publishers, 2005, p.345.

Chapter Seven: The Battle of Biosolids

169 *Bottles of cleaned sewage effluent* – Kevin R. Cowan, who runs the North Davis sewer district, hands out the sewer water bottles to visitors. They have also been handed out at meetings, where the response was 'tepid'. The bottles

actually contain drinking water. Lynn Arave, 'Sewer water in a bottle – yum!', *Deseret Morning News*, 14 October 2007.

170 *'A great rough sort of business'* – Throughout the late nineteenth century, plenty of sewage farming systems were proposed to the Metropolitan Board of London. In fact, wrote the anonymous author of one report, 'had the application of sewage to agricultural purposes been as easy and as profitable as some loud-talking and fluently writing people pretend, it would long since have formed the foundation of more than one joint-stock company'. *The agricultural value of the sewage of London examined in reference to the principle systems admitted to the Metropolitan Board of Works*, London: Edward Stanford & Co., 1865, pp.17–18, p.38.

172 *'Stench'* – On 11 July 2007, a Mogden resident called Fullalove sent the following complaint to Thames Water: 'FOUL YUK STINK STENCH I CANNOT BREATHE IT IS YOUR FAULT.' Two weeks later, his neighbour Jonathan Oatway wrote despairingly that Mogden was 'absolutely reeking again after the rain. It was stomach turning. We had to close all the windows but the smell lingered right throughout the house. I was moved to apologise to a guest we had visiting at the time. All very embarrassing considering the raw sewage nature of the odour. It's a joke living here, why should we live in fear of the stench?' Mogden Residents Action Group, http://www.mogden.org.uk. Whitley Treatment Works, meanwhile, inspired the Whitley Whiff. They have now been replaced by the entirely covered – and odour-free – Reading Wastewater Treatment Works, which cost £80 million to build.

173 *Synagro* – Biosolids are lucrative enough for Synagro to have been bought in 2007 by the powerful private equity group Carlyle for $772 million: Justin Baer, 'Carlyle raises Dollars 1bn for public works', *Financial Times*, 7 November 2007.

174 *'Bubbling with sub-surface gases'* – The fires burned with such intensity, two railway bridges spanning the river were nearly destroyed. The mayor of Cleveland, which the Cuyahoga runs through, said it was 'a terrible reflection on our city'. 'The Cities: The Price of Optimism', *Time*, 1 August 1969.

– *7 million dry tons of sludge* – John Heilprin and Kevin S. Vineys, 'Sewage-based fertiliser safety doubted', *Associated Press*, 6 March 2008.

175 *The worth of this lost wealth* – Halliday, op. cit., p.117.

– *'No better use for the excretion'* – Karl Marx, *Capital: A Critique of Political Economy*, vol. 3, Chicago: Charles H. Kerr & Company, 1909, p.120.

- *'Each farmer will turn on the tap'* – John Joseph Mechi, *How to farm profitably: The sayings and doings of Mr Alderman Mechi*, London: Routledge, 1860, p.352.
- *Vegetables highly sought after* – Gennevilliers sewage farm received 800 litres of Parisian sewage per second. Once irrigated, its previously barren soil became capable of producing 40,000 heads of cabbage, 60,000 artichokes, or 200,000 pounds of sugar beets per hectare. One Parisian perfumer raised peppermint. Reid, op. cit., pp.62–3.
- *The sewage farm at Pasadena* – 'By the end of the century sewage farms were operating in twenty locations including Salt Lake City and Los Angeles. Pasadena's sewage farm was renowned. The 300 acre farm supported a hundred pigs, an extensive tract of English walnuts, and an alfalfa crop largely dedicated to the municipality's working horses.' Jamie Benidickson, *The Culture of Flushing: A Social and Legal History of Sewage*, Vancouver: UBC Press, 2007, p.126.
- *US industry is estimated to use 100,000 chemicals* – Statement of Bruce Alberts, President, National Academy of Sciences, to U.S. Senate Committee on Environment and Public Works, 14 May 2002.http://www7.nationalacademies.org/ocga/testimony/Persistant_Organic_Pollutants.asp
- *1000 new chemicals being added every year* – Jonathan M. Harris, *A Survey of Sustainable Development*, Washington, D.C.: Island Press, 2001, p.290.

176 *Selling its sludge as fertilizer* – Milorganite was named by McIver and Son of Charleston, South Carolina, who entered a competition to name the new fertilizer in *National Fertilizer Magazine* in 1925. Milorganite stands for Milwaukee Organic Nitrogen. Its history is told at http://www.milorganite.com

- *Toxic Sludge* – John Stauber and Sheldon Rampton, *Toxic Sludge is Good for You: Lies, Damn Lies and the Public Relations Industry*, Monroe, Maine: Common Courage Press, 1995.

178 *Orange County* – Four US states have county-level grand juries. California's grand juries sit for a year and are mandated to issue investigative reports, usually into government services and institutions. In 2004, Orange County also issued reports on speaking English in Santa Ana, prisons, and whether improvements were needed in the county's animal shelters. Orange County Grand Jury, 'Does Anyone Want Orange County Sanitation District's 230,000 Tons of Biosolids?', 2003–4, at http://www.ocgrandjury.org/reports.asp#2004-2005

– *'Musty' or 'earthy'* – Ibid., pp.8–9.

181 *Twenty-five groups of pathogens* – 'Researchers Link Increased Risk Of Illness To Sewage Sludge Used As Fertiliser', University of Georgia, news release, 30 July 2002.

– *'An appropriate choice for communities'* – U.S. EPA, 'Standards for the Use or Disposal of Sewage Sludge; Final Agency Response to the National Research Council Report on Biosolids Applied to Land and the Results of EPA's Review of Existing Sewage Sludge Regulations', Federal Register: 31 December 2003 vol. 68, no. 250, pp. 75, 536.

182 *Thirty-eight months* – National Research Council, 'Biosolids Applied to Land: Advancing Standards and Practices,' Washington, D.C.: National Academy Press, 2002 (pre-publication copy), p.199.

– *'Unmonitorable, unregulatable and irremediable'* – Abby A. Rockefeller, 'Sewers, Sewage treatment, Sludge: Damage without end', *New Solutions*, vol. 12 (4) 2002, p.342.

183 *'EPA regional staff'* – U.S. EPA Office of Inspector General, 'Land Application of Biosolids' (2002-S-000004), 28 March 2002, p.6.

184 *Asthma, flu-like symptoms* – Ellen Z. Harrison and Summer Rayne Oakes, 'Investigation of Alleged Health Incidents associated with land application of sewage sludges', *New Solutions*, vol. 12 (4), 2002, p.387. An updated tally of health complaints is kept by the Cornell Institute for Waste Management at http://cwmi.css.cornell.edu/Sludge/incidents.htm

– *Chemical irritants in sludge* – David. L. Lewis and David K. Gattie, 'A High-level Disinfection Standard for Land-Applied Sewage Sludges (Biosolids)', *Environmental Health Perspectives* 112, 2004, p.127.

185 *Adverse human health effects* – National Research Council, op. cit., p.3.

– *'Big tomatoes'* – Tom Gibb, 'A terrible waste gets long look', *Pittsburgh Post-Gazette*, 11 June 2000.

– *Shayne Conner* – Laura Orlando, 'Sustainable Sanitation: A Global Health Challenge', *Dollars & Sense*, May–June 2001.

– *Sludge was behind it* – 'Sewage Fertiliser under Fire', CBS News, 29 October 2003, http://www.cbsnews.com/stories/2003/10/29/eveningnews/main580816.shtml

186 *The company has yet to answer* – In response to questions about why the case had been settled, Synagro Public Relations Director Lorrie Loder replied, 'I do not have that information available to me. I am generally aware that cases are

frequently settled as a matter of expediency.' Personal communication with Lorrie Loder, Synagro, April 2008.

– *Pesticides were well within legal limits* – Douglas Fischer, 'Chemical: Mixtures more toxic than their parts', *Oakland Tribune*, 24 January 2006.

– *'I can't answer it's not safe'* – CBS News, op. cit.

– *Lead is down; mercury is up* – A small-scale study carried out by the *Oakland Bay Tribune* found that an average family contained hundreds of chemicals in their bodies, including 'plasticizers (phthalates), combustion products (polycyclic aromatic hydrocarbon metabolites and dioxins/furans), chlorinated hydrocarbon pesticides (DDT, hexachlorobenzene, lindane, chlordane, heptachlor epoxide), PCBs, organophosphorus insecticide metabolites, other pesticides (various herbicides, including atrazine, 2,4-D/2,4,5-T and pentachlorophenol), tobacco smoke indicators (cotinine) and phytoestrogens'. The family's two-year-old son had 838 parts per billion of PBDEs – the flame retardants Dr Rob Hale is examining in sludge – in his blood, although the world average is 38. Douglas Fischer, 'What's in you?' *Oakland Bay Tribune*, 3 August 2005. Centers for Disease Control and Prevention (CDC), 'Third National Report on Human Exposure to Environmental Chemicals', 2005.

188 *'Is the risk acceptable?'* – Ellen Z. Harrison, Murray B. McBride and David R. Bouldin, 'The Case for Caution', Cornell Waste Management Institute Working Paper, February 1999, p.3.

– *Sanjour wrote another memo* – Collected papers of William Sanjour, http://pwp.lincs.net/sanjour/Default.htm

– *'Didn't think [the Part 503 rules] passed scientific muster'* – C. Snyder, 'The Dirty Work of Promoting "Recycling" of America's Sewage Sludge', *International Journal of Occupational Environmental Health*, 2005; 11, p.417.

189 *The immuno-compromised* – 'Because data are sparse on what constitutes an infective dose, it is prudent public health practice to minimise workers [sic] contact with Class B biosolids and soil or dusts containing Class B biosolids during production and application, and at land application sites during the period when public access is restricted.' National Institute of Occupational Safety and Health and Centers for Disease Control and Prevention, 'Guidance for Controlling Potential Risks to Workers Exposed to Class B Biosolids', July 2002, p.2.

– *Lubricants used in dental devices* – David L. Lewis and Max Arens, 'Resistance of microorganisms to disinfection in dental and medical devices', *Nature Medicine* 1, 1995, pp.956–8.

– *Scientific rules pushed through* – David L. Lewis, 'EPA science: casualty of election politics', *Nature* 381, 27 June 1996.

– *Unreliable and fraudulent data* – McElmurray v. United States Department of Agriculture, United States District Court for the Southern District of Georgia, Southern Division, Case 1:05-cv-00159-AAA-WLB, judgement by Judge Anthony Alaimo, filed 2 February 2008, p.39.

– *Simply 'not a good idea'* – Jessica Leeder, 'Human Fertilizer poses cancer risk: Study', *National Post*, 31 July 2002; McElmurray v. USDA, op. cit., p.41.

– *Questioned his credibility* – Josh Harkinson, 'Wretched Excess', *Houston Press*, 31 March 2005.

– *EPA's Science and Technology Achievement Award* – personal communication with Maggie Breville, U.S. EPA, March 2008.

190 *The EPA's treatment of Lewis* – Caroline Snyder, 'EPA wants scientist out for publishing papers critical of sludge rule', National Treasury Employees Union Chapter 280 Newsletter, July 2002, vol. 18, no. 5.

– *The safest science-based alternative* – Letter from Albert Gray, Water Environment Foundation, to the *Washington Post*, 7 August 2001, available at http://www.biosolids.org/news.asp?id=1219

191 *'I should think myself a madman'* – Not all Barking residents proved to be good witnesses for the vicar's cause. Mr Frederick Powell, lighterman, when asked if his fellow residents often complained of the smell of sewage in the creek, said, 'I never heard persons who were sitting in the Ship say, "How beastly the London sewage smells! Let us drink up and go."' Robert Rawlinson, *Report upon Inquiry as to the truth or otherwise of certain allegations contained in a memorial from the vicar and other inhabitants of Barking, in the County of Essex, calling attention to the pollution of the River Thames by the Discharge of Sewage through the Northern Main Outfall Sewer of the Metropolitan Board of Works*, London: George Edward Eyre and William Spottiswoode for Her Majesty's Stationery Office, 1870, pp.4, 99.

– *'Entirely imaginary and contrary to the fact'* – Ibid., p.42.

– *Vomited copiously* – 'There can be no doubt,' wrote *The Times* in one of many leaders about the disaster, 'that at the period when the collision occurred, the Metropolitan sewers – that for the north side at Barking and that for the south side at Belvedere – were pouring forth their daily contribution of millions of gallons of water loaded with all the filth of a great city.' One survivor told a subsequent inquiry that death may have been 'due in many cases to the poisonous state of the water'; another said that 'both for taste and smell it was something he could hardly describe'. Three inquiries blamed first the *Bywell Castle* dredger, then the *Princess Alice*, then both ships. Among the 631 victims, most of whom could not swim or were hampered by being children or wearing copious petticoats, were 'four children of Mr and Mrs Davies, 281, Burdett-Road, Limehouse, [and] Arthur Kiddell, a little boy, who was visiting them'. *The Times*, 18 and 19 September 1878. A detailed account of the disaster has been posted by the Thames Police Museum at http://www.thamespolicemuseum.org

– *'A natural contaminant'* – Devra Lee Davis, *The Secret History of the War on Cancer*, New York: Basic Books, 2007, pp.78–9.

– *Judges in Kentucky, California and Oregon* – Penland v. Redwood Sanitary Sewer District 1998 via http://www.publications.ojd.state.or.us/A90247.htm; Jamie Manfuso and Scott Carroll, 'Sludge ban starts fight across nation', *Sarasota Herald-Tribune*, 14 July 2002.

– *120 times those allowed in drinking water* – Nonetheless, after the Boyce Dairy informed the EPA that it found thallium and other heavy metals in its milk, the dairy was still allowed to sell the milk for human consumption. Heilprin and Vineys, op. cit.

– *Unreliable, incomplete and in some cases fudged* – McElmurray v. United States Department of Agriculture, op. cit., p.15.

192 *National Farmers' Union policy* – Policy of the National Farmers' Union, enacted by delegates to the 104th anniversary convention, Denver, Colorado, 3–6 March 2006, p.69.

– *Switzerland . . . banned the practice* – *'La fin des boues d'épuration dans l'agriculture'*, Office Fédéral de l'Environnement, news release, 13 May 2002.

Chapter Eight: Open Defecation Free India

196 *Every day, 200,000 tonnes of faeces* – United Nations Sustainable Development Division, 'Nirmal Gram Puraskar: Fiscal rewards to zero open defecation in rural villages in India', case study 2003–5, available from http://webapps01.un.org/dsd/caseStudy/public/Welcome.do

– *155,000 truckloads* – Darryl D'Monte, 'A bottom-up approach to sanitation',

Infochange India, October 2006, http://www.infochangeindia.org/features389.jsp

– *Sitting on their haunches* – Singh, op. cit., p.5.

– *They did it beside train tracks* – 'Indians defecate everywhere. They defecate, mostly, beside the railway tracks. But they also defecate on the beaches; they defecate on the hills; they defecate on the river banks; they defecate on the streets; they never look for cover.' V.S. Naipaul, *Area of Darkness*, London: Picador, 1964, reprinted 2002, p.70.

– *'Scores of bare bottoms'* – Chander Suta Dogra, 'Whole Lota Love', *Outlook*, 24 July 2006.

197 *Blocking natural body functions* – UNICEF, 'Meeting the MDG Water and Sanitation Target: A mid-term assessment of progress', undated, http://www.unicef.org/wes/mdgreport/

198 *1 billion people are carrying hookworm* – Because hookworm is generally tolerated and usually doesn't kill, exact figures about its prevalence are difficult to calculate. A 2003 review of the existing literature and figures put hookworm prevalence at 700–800 million and ascariasis at 1.2 billion, with 50 per cent of infections occurring in China. Nilanthi R. de Silva et al, 'Soil-transmitted helminth infections: updating the global picture', *Trends in Parasitology*, vol. 19, no. 12, December 2003, p.547.

– *The number of infections that faeces can transmit* – Richard G. Feachem, David J. Bradley, Hemda Garelick, D. Duncan Mara, *Sanitation and Disease: Health aspects of excreta and wastewater management*, Chichester: John Wiley & Sons for The World Bank, 1983, p.3.

– *Typhoid, scabies and botulism* – Ibid., pp.9–12; World Health Organization, Water and sanitation diseases fact sheets, from http://www.who.int/water_sanitation_health/diseases/diseasefact/en/index.html

– *F-diagram* – The F-diagram of faecal-oral transmission routes was devised in 1958 by E. G. Walter and J. N. Lanoix, 'The Handwashing Handbook', Washington, DC: The World Bank, 2005, p.10.

– *Nearly 800 million Indians* – 'Oxfam: Millions in South Asia lack vital services', *Associated Press*, 19 October 2006; UNDP, op. cit., p.36.

– *7.4 million more people* – WaterAid, 'Drinking Water and Sanitation Status in India,' 2005, p.25.

199 *Unused, misused or ignored* – A.J. Robinson, 'Scaling up Rural Sanitation in South Asia: Lessons learned from Bangladesh, India and Pakistan', New Delhi: Water and Sanitation Program-South Asia, 2005, p.42.

200 *Balancing a toilet on his head* – Darryl D'Monte, 'Slow progress towards sanitation', *India Together*, December 2004.

204 *Improved health never came into it* – Other reasons for wanting a latrine included not feeling embarrassed by having to indicate to important visitors where they should go to defecate; ensuring themselves a good place in the afterlife by leaving a durable legacy for their descendants; and fearing supernatural illnesses caused by smelling other people's faeces. Marion W. Jenkins and Val Curtis, 'Achieving the "good life": Why some people want latrines in rural Benin', *Social Science & Medicine* 61 (2005), pp.2450–1.

205 *Wants, not needs* – Val Curtis, 'Hygiene and Sanitation: Dirt, Disgust and Desire', Public Hygiene Lecture at the School of Oriental and Asian Studies, London, 9 October 2006.

208 *Bringing in the army* – Oxfam, 'Guidelines for Public Health Promotion in Emergencies', May 2001, p.8.

– *361 villages* – Personal communication with Joe Madiath, October 2007.

210 *Only $11* – Robinson, op. cit., p.85.

– *Rural Sanitary Marts* – WSSCC, *Listening*, p.27.

211 *'Not the GDP or Sensex'* – Anjali Puri, 'Second Nature,' *Outlook*, 24 July 2006.

– *The number of India's toiletless* – Robinson, op. cit., p.85.

– *10,000 villages applied* – Government of India, Department of Drinking Water Supply, http://nirmalgrampuraskar.nic.in

– *Anyone running for local office* – Nirmala Ganapathy, 'No toilet at home? Don't contest panchayat polls', *Indian Express*, 4 November 2005.

212 *'A key factor which triggers mobilization'* – Kar offers other tips for trainee triggerers, including, 'On the transect walk, draw attention to the flies on the shit, and the chickens pecking and eating the shit. Ask how often there are flies on their, or their children's, food, and whether they like to eat this kind of local chicken.' Kamal Kar, 'A practical guide to triggering Community-Led Total Sanitation (CLTS)', Institute of Development Studies, University of Sussex, 2005, p.6.

213 *'Where the hell does it all go?'* – WSSCC, *Listening*, p.40.

– *'Onto the flies'* – Ibid.

– *A huge online poll* – Via a BBC website (www.bbc.co.uk/science/humanbody/mind/disgust), visitors were invited to rate a series of twenty paired images according to how disgusting they were. Seven of the pairs deliberately placed a disease-related image next to a similar image that wouldn't cause disease. A cut hair can carry ringworm; the hair attached to someone's head is safe. Over four months, 77,000 people from 165 countries took part. The plate of yellow goo – supposed to look like bodily fluids – rated as sixty-one per cent more disgusting than its paired image of blue goo. A final question asked respondents to rate who they would least want to share a toothbrush with. The postman rated as most disgusting (59.4 per cent); a partner as the least disgusting (1.8 per cent). These findings correspond to the fact that the less familiar the person is, the more likely it is that he/she will carry possibly transmittable pathogens. Disgust denoted threat. Val Curtis, Robert Aunger and Tamar Babie, 'Evidence that disgust evolved to protect from risk of disease', *Proceedings of the Royal Society* B (Suppl.) 271, 2004, pp.131–3.

214 *Their own babies' faeces* – In two studies, mothers were first asked to fill in a questionnaire about how they felt about changing their own baby's faeces-soiled diaper. In a second study, they were presented with samples of their own baby's dirty diaper and diapers from other infants. Evidence showed that mothers found their own baby's diaper smell less disgusting, even when labelling didn't make it clear whose diaper was whose. Case, Repacholi, and Stevenson, op. cit.

– *Dirty because it is out of place* – 'Dirt then, is never an isolated, unique event. Where there is dirt, there is system. Dirt is the by-product of a systematic ordering and classification of matter, in so far as ordering involves rejecting inappropriate elements.' Mary Douglas, *Purity and Danger: An analysis of concepts of pollution and taboo*, London: Routledge, 2006, p.44.

– '*The ultimate* lèse-majesté' – John Berger, 'Muck and its entanglements', *Harper's Magazine*, May 1989, pp.60–1.

215 *The Bangladesh programme* – An interesting account of Mosmoil's story is told in a film by the World Bank, narrated by the Indian actor Roshan Seth, which re-enacts the CLTS process using the real villagers of Mosmoil. *Awakening: The Story of Total Sanitation in Bangladesh*, available online at http://www.wsp.org/filez/video/4162007110732_AwakeningPart1.wmv

216 *'Bare bottoms doing what they must'* – Gourisankar Ghosh, formerly chief of the WSSCC, told me that the reason former Indian Prime Minister Rajiv Gandhi became interested in sanitation was because he took a Delhi–Bombay fast train one morning and happened to look out of the window. V.S. Naipaul, who famously called India 'the turd world', wrote that open defecation was so endemic because 'it is said that the peasant, Muslim

or Hindu, suffers from claustrophobia if he has to use an enclosed latrine'. Naipaul, op. cit., p.70.

218 *A thousand tigers* – WSSCC, *Listening*, p.41

222 *'Gifting latrine slabs to brides'* – Water and Sanitation Program, *Awakening: The Story of Total Sanitation in Bangladesh*, op. cit.

Chapter Nine: In the Cities

226 *'The hydraulic city'* – Matthew Gandy, 'Water, Sanitation and the Modern City: Colonial and Post-colonial Experiences in Lagos and Mumbai', Human Development Report Occasional Paper, 2006, p.4.

– *'Operational definition' [of slums]* – United Nations Human Settlement Programme (UN-HABITAT), 'The Challenge of Slums: Global Report on Human Settlements', London and Sterling, VA: Earthscan, 2003, p.12.

228 *Now an urban species* – The United Nations Population Division expected the milestone to have been reached in 2005; Worldwatch's *State of the World 2007* report concluded it would happen in 2008. David Whitehouse, 'Half of humanity set to go urban', BBC News, 19 May 2005; 'State of the World 2007: Notable Trends', Worldwatch, news release, 10 January 2007.

– *Nearly a billion slum-dwellers* – UN-HABITAT, op. cit., p.xxv.

– *Africa's slum-dwelling population* – John Vidal, 'Cities are now the frontline of poverty', *Guardian*, 2 February 2005.

– *Kenya's population growth* – Mike Davis, *Planet of Slums*, London: Verso, 2006, pp.18, 24.

– *One hundred thousand people move to slums* – 'The growth of cities: monsters stir', IRIN (UN Office for the Coordination of Humanitarian Affairs), 15 March 2008.

229 *$1.4 billion a year* – Dan McDougall, 'Waste not, want not in the £700m a year slum', *Observer*, 4 March 2007.

232 *Sewers . . . cost five times more* – David Nilsson, 'A heritage of unsustainability? Reviewing the origin of the large-scale water and sanitation system in Kampala, Uganda', *Environment & Urbanization*, vol. 18, 2006, p.380.

– *Laid in shallower trenches* – Duncan Mara, 'Health and Sanitation in the Developing World', paper delivered at the World Toilet Summit, Singapore, 19–21 November 2001, http://www.personal.leeds.ac.uk/~cen6ddm/papers 2000-2005.html

– *The authority generally provided trunk sewers* – The OPP method also saves money. Residents invested 90 million rupees ($1.4 million) to build their neighbourhood sanitation systems; it's estimated the same level of service would have cost local government $10.5 million. Arif Hasan, 'Orangi Pilot Project: the expansion of work beyond Orangi and the mapping of informal settlements and infrastructure', *Environment & Urbanization* vol. 18 (2), 2006, pp.451–480.

– *Repeated in forty-two other Karachi slums* – UNDP, op. cit., p.121.

236 *SPARC's business is booming* – A precise figure of SPARC's reach is difficult, but the projects in Mumbai and Pune already serve a quarter of a million people. Personal communication with David Satterthwaite, International Institute for Environment and Development, London, April 2008.

– *'Foolish to pass from one distortion'* – Jeremy Seabrook, quoted in Davis, op. cit., p.70.

237 *ninety-six per cent of Tanzanians* – WHO/UNICEF, 'Meeting the MDG Drinking Water and Sanitation Target', 2006, p.43.

238 *'Faecal contamination of users'* – David Satterthwaite and Gordon McGranahan, 'Overview of the Global Sanitation Problem', Human Development Report Office Occasional Paper No. 27, New York: UNDP, 2006, p.8.

239 *It was cholera that made headlines* – Steven Shapin, 'Sick City', *New Yorker*, 6 November 2006.

240 *Melbourne uses waste-stabilization ponds* – Melbourne Water operates the Western Treatment Plant at Werribee, which treats 485 million litres a day, provided by 1.6 million people. Further information can be found on the extensive website of Professor Duncan Mara, a world authority on ponds. See http://www.personal.leeds.ac.uk/~cen6ddm/

242 *Less than half a cent per person* – Marion W. Jenkins and Steven Sugden, 'Rethinking Sanitation: Lessons and Innovation for Sustainability and Success in the New Millennium', Human Development Report Office Occasional Paper No. 27, New York: UNDP, 2006, p.25.

245 *US patent 6,242,489* – available at http://www.uspto.gov/patft/

– *'Or as a taggant'* – The Sunshine Project, Backgrounder Series #8, July 2001, http://www.sunshine-project.org/publications/bk/bk8en.html

247 *John Snow . . . never patented* – Peter Vinten-Johansen, *Cholera, Chloroform and the Science of Medicine: A Life of John Snow*, New York: Oxford University Press, 2003, p.113.

248 *'Compact disc players or mobile phones'* – Jenkins and Sugden, op. cit., p.13.
249 *A toilet shop was a great idea* – Steadman & Associates, 'Social Marketing for Urban Sanitation Pilot Survey in Keko Mwanga B', 2003.
251 *Punishing running costs* – Personal communication with Steven Sugden, April 2008.

Chapter Ten: The End

255 *$23.4 million . . . toilet* – 'An Astronomical Potty', *Time*, 25 January 1993.
256 *Its precious fabrics* – Tom McNichol, 'The Big Gulp', *Wired*, August 2005.
257 *$40,000 to transport each gallon* – Ibid.
– *A Russian second-hand . . . toilet* – 'NASA buys $19 million Russian toilet system for international space station', *International Herald Tribune*, 5 July 2007.
258 *Chronically short of water* – All water facts in this paragraph are taken from Robin Clarke and Jannet King, *The Atlas of Water: Mapping the World's Most Critical Resource*, London: Earthscan, 2004, p.19.
– *One cubic metre of wastewater* – Ibid., p.11.
– *'Bolted onto every sewage plant'* – 'From the toilet to the tap', Australian Broadcasting Corporation, 9 November 2006, transcript from http://www.abc.net.au/catalyst/stories/s1785041.htm
259 *Upper Occoquan Sewage Authority* – Personal communication with Robert Bastian, US EPA Office of Wastewater Management, April 2007.
– *The French minister of the environment* – *'Eau: une pub source de polémique'*, TF1, 17 January 2007, transcript from http://tf1.lci.fr/infos/sciences/environnement/0,,3381854,00-eau-pub-source-polemique-.html
– *$2.17 billion-worth of only one brand* – Johnny Davis, 'Would madam care to taste the cloud juice?' *Guardian*, 2 December 2007.
– *Out of an ordinary tap* – 'Pepsi says Aquafina is tap water', CNN, 27 July 2007, http://money.cnn.com/2007/07/27/news/companies/pepsi_coke/. A four-year study by the Natural Resources Defense Council into America's bottled-water industry found that a quarter of bottled 'mineral' waters are in fact municipal tap water. Erik D. Olson, 'Bottled Water: Pure Drink or Pure Hype?' Natural Resources Defense Council, February 1999, available at http://www.nrdc.org/water/drinking/bw/bwinx.asp
– *'The consumer just doesn't seem to care'* – Davis, op. cit.
263 *'Leapfrog over the dinosaur technologies'* – Stephen Dale, 'Regenerative Solutions

for Managing Community-generated Organic Waste', *Reports* (magazine of the International Development Research Center), 4 February 2000, available from http://www.idsc.ca/en/ev-5574-201-1-DO_TOPIC.html

264 *11.5 watts* – Graham Lawton, 'Pee-cycling', *New Scientist*, 20 December 2006.

– *'A carbon impact'* – 'Future water: The Government's water strategy for England', HM Government, 2008, p.6.

– *'Can't be good'* – Jeff Donn, Martha Mendoza and Justin Pritchard, 'AP Probe finds Drugs in Drinking Water', *Associated Press*, 9 March 2008.

– *David Stuckey* – 'Submerged Anaerobic Membrane Bioreactor Awarded', *Water and Wastewater Newsletter*, vol. 10, no. 329, 17 March 2008.

265 *One toilet per second* – Personal communication with Arno Rosemarin, Swedish Environment Institute, November 2006.

268 *'Anything but stable'* – WHO, 'World Health Report 2007: A Safer Future: Global Public Health Security in the 21st Century', Geneva: WHO Press, 2007, p.vi.

– *Eleven per cent of those died* – Ibid., p.39.

– *A disease of prosperous urban centres* – Ibid., p.40.

269 *Planners and Searchers* – William Easterly, *The White Man's Burden: Why the West's Efforts to Aid the Rest Have Done So Much Ill and So Little Good*, Oxford: Oxford University Press, 2007.

– *The excuse of ignorance* – Harold Farnsworth Gray, 'Sewerage in Ancient and Modern Times', *Sewage Works Journal*, vol. 12, no. 5, September 1940, p.946.

Further Reading

Barnes, David S., *The Great Stink of Paris and the Nineteenth-Century Struggle Against Filth and Germs*, Baltimore, MD: Johns Hopkins University Press, 2006.

Black, Maggie and Fawcett, Ben, *The Last Taboo: Opening the Door on the Global Sanitation Crisis*, London: Earthscan, 2008.

Chadwick, Edwin, *Commentaries on the Report of the Royal Commission on Metropolitan Sewage Discharge and on the Combined and Separate Systems of Town Drainage*, London: Longmans & Co., 1885.

Cockayne, Emily, *Hubbub: Filth, Noise, and Stench in England, 1600–1770*, New Haven, CT, London: Yale University Press, 2007.

Dundes, Alan, *Life is like a Chicken Coop Ladder: A Study of German National Character through Folklore*, Detroit: Wayne University State Press, 1989.

Dyke, George Vaughan, *John Lawes of Rothamsted, Pioneer of Science, Farming and Industry*, Harpenden, Herts: Hoos, 1993.

Eveleigh, David, *Bogs, Baths and Basins: The Story of Domestic Sanitation*, Stroud, Gloucestershire: Sutton, 2006.

Gavin, Hector, *Sanitary Ramblings: Being Sketches and Illustrations of Bethnal Green; A Type of the Condition of the Metropolis and other Large Towns*, London: John Churchill, 1848.

Glassford, Charles F.O., *Town Excreta: Its Utilization. London Sewage, Shall it be Wasted? Or Economised?* London, 1858.

Horan, Julie L., *Sitting Pretty, An Uninhibited History of the Toilet*, London: Robson Books, 1998.

Hoy, Suellen, *Chasing Dirt: The American Pursuit of Cleanliness*, Oxford University Press, 1995.

Inglis, David, *A Sociological History of Excretory Experience: Defecatory Manners and Toiletry Techniques*, Lewiston, NY, Lampeter: Edwin Mellen Press, 2000.

Jenkins, Joseph, *The Humanure Handbook: A Guide to Composting Human Manure*, White River Junction, VT: Chelsea Green Publishing, 2005.

Kilroy, Roger, *The Compleat Loo: A Lavatorial Miscellany*, London: Gollancz, 1984.

Lewin, Ralph A., *Merde: Excursions in Scientific, Cultural and Socio-Historical Coprology*, New York: Random House, 1999.

Markham, Len, *Yorkshire Privies: A Nostalgic Trip down the Garden Path*, Newbury, Berkshire: Countryside Books, 1996.

McLaughlin, Terence, *Coprophilia, or a Speck of Dirt*, Cassell, 1971.

Miller, Ian William, *Anatomy of Disgust*, Cambridge, MA, London, Harvard University Press, 1997.

Ogle, Maureen, *All the Modern Conveniences: American Household Plumbing, 1840–1890*, Baltimore: The Johns Hopkins University Press, 1996.

Palmer, Roy, *The Water Closet, A New History*, Newton Abbot: David & Charles, 1973.

Pathak, Bindeshwar, *Road to Freedom: A Sociological Study on the*

Abolition of Scavenging in India, New Delhi: Motilal Banarsidass Publishers, 1998.

Pomfret, John, *Chinese Lessons: Five Classmates and the Story of New China*, New York: Henry Holt, 1996.

Praeger, Dave, *The Poop Report: How America is Shaped by its Grossest National Product*, Los Angeles: Feral House, 2007.

Pudney, John, *The Smallest Room*, London: Michael Joseph, 1954.

Sabbath, Dan & Hall, Mandell, *End Product: The First Taboo*, New York: Urizen Books, 1977.

Sale, Charles, *The Specialist*, London: Charles Putnam and Company, 1956.

Smith, Stephen, *Underground London: Travels Beneath the City Streets*, London: Little, Brown, 2004.

Stanbridge, H.H., *History of Sewage Treatment in Britain*, Maidstone: Institute of Water Pollution Control, 1976–1977.

Toscani, Oliviero, *Cacas*, Köln: Taschen, 1988.

Thekaekara, Mari Marcel, *Endless Filth: The Saga of the Bhangis*, London: Zed Books, 2003.

Filmography

Boys from the Brown Stuff, dir. David Clews (Blast Films for BBC 2, 2007)

Faecal Attraction, dir. Pradip Saha (Centre for Science and the Environment, 2006)

Kenny: A Knight in Shining Overalls, dir. Clayton Jacobson (Thunderbox Films, 2006)

Q2P, dir. Paromita Vohra (Devi Pictures, 2006)

Sludge Diet, dir. Mario Desmarais (Productions Thalie, 2006)

The Casteless, dir. Jens Pedersen (Danish Broadcasting Corporation, 2006)

The Snowball Effect, dir. Klee Benally (Indigenous Action Media, 2005)

Waste not Waste, dir. Joshka Wessels (Television for the Environment, 2008)

Index